# Introduction to Quantum Chemistry

# Introduction to Quantum Chemistry

Clifford E. Dykstra

*Department of Chemistry*
*Indiana University-Purdue University*
*at Indianapolis*

Prentice Hall, Englewood Cliffs, New Jersey 07632

*The Library of Congress has cataloged the Standard Edition of this work as follows:*

**Library of Congress Cataloging-in-Publication Data**

Dykstra, Clifford E.
Quantum chemistry and molecular spectroscopy / Clifford E. Dykstra.
p. cm.
Includes bibliographical refereces and index.
ISBN 0-13-747312-5
1. Quantum chemistry. 2. Molecular spectroscopy. I. Title.
QD462.D95 1991
541.2'8—dc20

Editor-in Chief: Tim Bozik
Production Editior: Tom Aloisi
Cover design: Violet Lake
Production Coordinator: Trudy Pisciotti

A Paramount Communications Company
Englewood Cliffs, New Jersey 07632



Printed in the United States of America
10 9 8 7 6 5 4 3 2 1

ISBN 0-13-701293-4

Prentice-Hall International (UK) Limited, *London*
Prentice-Hall of Australia Pty. Limited, *Sydney*
Prentice-Hall Canada Inc., *Toronto*
Prentice-Hall Hispanoamericana, S.A., *Mexico*
Prentice-Hall of India Private Limited, *New Delhi*
Prentice-Hall Japan, Inc., *Tokyo*
Simon & Schuster Asia Pte. Ltd., *Singapore*
Editora Prentice-Hall do Brazil, Ltda., *Rio de Janeiro*

# Contents

# *Preface*

This text was developed for an introductory course in the application of quantum mechanics to chemistry. It presents both formal elements of quantum theory and the analysis for specific problems in infrared spectroscopy, electronic structure, magnetic resonance, and other areas. The text is intended for use in a junior or senior level undergraduate course of up to fifteen weeks (45 lecture hours).

Prerequisite work should include solid preparation in basic calculus (i.e., one year of coursework). In most places in the text, details of the mathematical steps are given fully, and so math preparation beyond calculus is not necessary. However, two chapter sections require some familiarity with linear algebra. These two sections are designated with a ✭ symbol in the contents. Appendix I provides background material in matrix algebra that is needed for these sections; however, the two sections may be excluded without complicating the development of later material.

An aim of this text is to provide a concise yet thorough introduction to quantum chemistry and the basis for molecular spectroscopy. Differences with traditional presentations of this material represent my attempt to streamline the subject and to make the presentation as up-to-date as possible. There is an early presentation of the harmonic oscillator, for instance, because it then serves as a primary model problem for several quantum concepts and for diatomic infrared spectroscopy. Descriptive information about certain advanced topics is given from time to time in order to acquaint students with what lies ahead.

The experimental basis of quantum chemistry and molecular structure is a crucial aspect of the subject. The aim of the text has been to develop quantum mechanics to the point of predicting and interpeting spectra. It is intended that students learn the connection between quantum mechanical analysis and the measurements they may make in a laboratory, but topics related to instrumentation are not considered.

The exercises at the end of each chapter bring out essential points of the material. Sometimes they amount to completing or extending a problem that had been discussed, and generally, they pose a concise challenge to a student's understanding. Notions that emerge from reading the text will be strengthened by completing the exercises. At the end of the text, solutions are given for a number of the problems.

I should like to thank Dr. Joseph Augspurger for preparing a number of the solutions to the chapter exercises and for preparing several figures.

*C. E. D.*

Indianapolis

1993

# Introduction to Quantum Chemistry

# Chapter 1

# Introduction

*The most detailed pictureof molecules is that provided by quantum mechanics. Quantum chemistry refers to the aspects of quantum theory that are of particular interest in chemistry. The development of quantum chemistry has occurred through a strong interplay of experimental investigation, especially molecular spectroscopy, and theoretical analysis.*

## 1.1 CLASSICAL AND QUANTUM PHYSICS

The most detailed view of atoms and molecules is one that reveals their constituents: electrons, neutrons, and protons. These constituents are particles, very small entities that have mass. They are so small and so light that they are beyond the limits of our own senses and experience. Not surprisingly, the mechanics of systems of such particles are also outside everyday experience. Even so, there is a correspondence between our macroscopic world and the subatomic world, and we will analyze systems of particles in both worlds. The picture in the macroscopic world is generally referred to as a *classical* picture. It has been established by human perception and observation. The picture of small, light particles is termed a *quantum* picture because quantization of energy – the partitioning of energy into discrete chunks – and of other things is the distinction between this world and the macroscopic world. Because the discrete chunks are so tiny, they appear to constitute something continuous in the macroscopic world, and so there is at least this connection with our usual experience.

The modern atomic theory of matter is almost two centuries old. It was in the early 1800's that Dalton's work brought the first evidence that matter is not continuously divisible and that there is some fundamental type of particle, the atom. The line of thought that began with the atomic theory of matter took its next major step in the early 1900's when experiments pointed to the existence of subatomic particles. In the decades since, it has become clear that there are many smaller particles. The important notion is that matter is composed of discrete building blocks (particles) rather than continuous materials.

Einstein's special theory of relativity in 1905 connected the property of mass with energy. Perhaps in a vague way this connection makes it less surprising that scientists in the early 1900's were finding that many strange observations could be explained if energy came in discrete chunks. In other words, it is not only matter but also energy that comes in discrete building blocks in the tiny world of atoms. The problems that led to the hypothesis of quantization of energy involved the absorption spectrum of hydrogen, the photoelectric effect, and the temperature dependence of the heat capacities of solids. One after another, unexplained phenomena were explained by a quantum hypothesis, and this hypothesis eventually grew into what we now refer to as *quantum mechanics*. These early developments truly mark the start of a major scientific revolution.

## 1.2 THEORY AND EXPERIMENT

The pursuit of understanding in most branches of science is a process of observation and analysis. In chemical physics, laboratory experiments are the means for observation, which is to say that experiment is the means for probing and measuring molecular systems. The analysis of the data, though, may be carried out for different reasons. For one, the analysis might use the data with some generally accepted theory so as to deduce some useful quantity that is not directly measurable. We will find, for instance, that bond lengths are not measured the way an object's length is measured in our macroscopic world; instead they are often determined on the basis of measurements of energy changes in molecules. In this way, the established physical understanding provides the means for utilizing experimental information in examining molecular systems. So, quantum mechanical theories and experiments must go hand in hand.

Another purpose for carrying out an experiment and analyzing the data is to test one notion, concept, model, hypothesis, or theory, or possibly to select from among several competing theories. If the data do not conform to what is anticipated by one theory, then its validity has been challenged. In such a circumstance, one may devise a new notion, concept, or theory to better fit the data, or possibly reject one concept in favor of another. Whatever new idea emerges is then rightly tested in still further experiments. We will find that many problems are analyzed with approximations or idealizations that make the mathematics less complicated or that offer a more discernible physical picture. Then, experimental testing offers a validation or rejection of the approximation.

Throughout science there is a serious interplay between observation and understanding, or between experiment and theory. In quantum chemistry, one goal is a physical understanding of how molecular systems exist and behave, and ultimately that is embodied in theories that at some point have been well tested by experiments. In many respects, a textbook presentation of our understanding of quantum chemistry is a presentation of theories, and yet, experiments are very much a part of the story. We can not properly explain our best physical picture of molecular behavior without knowing the means of observation (experiment). The direction for this text is to present our basic quantum mechanical picture of molecules (theories) integrated with discussion of certain essential experiments.

## Bibliography

1. A. J. Ihde, *The Development of Modern Chemistry* (Harper & Row, New York, 1964).
2. T. S. Kuhn, *The Structure of Scientific Revolutions* (University of Chicago Press, Chicago, 1970). This is an insightful essay on what constitutes a scientific revolution and how the thinking of scientists changes as a revolution unfolds.
3. T. S. Kuhn, *Black-Body Theory and the Quantum Discontinuity, 1894-1912* (Oxford University Press, New York, 1978). This provides a fascinating and critical account of the development of the origins of quantum mechanics. In particular, it challenges the account of many texts about the origin of the quantum hypothesis.

# Chapter 2

# Classical Mechanics

*Mechanics is the analysis of the motions of objects. In this chapter, we consider the mechanics of particles in our macroscopic world. This classical description will be connected to the very different description of microscopic particles that will be presented in later chapters. Classical mechanics allows for continuous variability in the momenta and energies of particles, whereas at the quantum or microsocopic level, this does not always happen.*

## 2.1 EQUATIONS OF MOTION

Differential equations that prescribe the motion of a particle or system of particles are *equations of motion*. In the mechanics of Isaac Newton, the equations of motion are one of Newton's laws: The total force acting on an object equals the mass of the object times the time rate of change of the object's velocity (i.e., the acceleration). Letting $\vec{F}$ be the vector of force, m the mass of the object, and $\vec{a}$ the acceleration vector, then the relation

$$\vec{F} = m\,\vec{a} \tag{2-1}$$

is an equation of motion. If one finds or knows $\vec{F}$, this equation can be used to find out everything about the object's motion in space and time. Solution of this differential equation will yield the position vector as a function of time, i.e., $\vec{r}(t)$. This is a *classical mechanics* picture; nothing appears quantized.

Newton's formulation is not the only way in which classical equations of motion can be found. Lagrange and Hamilton developed other means, and it is the formulation of Hamilton that is the most useful framework for developing the mechanics of quantum systems. It is important to realize that Newtonian, Lagrangian, and Hamiltonian mechanics offer equivalent descriptions of classical systems. The $\vec{r}(t)$ found with one formulation will be the same as that found with another, even if the route to that function appears different.

At this point, we shall restrict attention to the mechanics of systems of point-mass particles. This means that particles are assumed to have no volume. Their mass is in an infinitesimally small region of space, a point. And until later, we will restrict attention to *conservative* systems, which are those for which the potential energy has no explicit time dependence. Of course, these restrictions are not an aspect of a particular mechanical formulation; they are used as a convenience that is in keeping with the types of systems of most immediate interest.

The classical kinetic energy, T, of any particle is the square of the momentum divided by twice the particle's mass. Momentum is a vector: In a Cartesian space, there is an x-component, a y-component, and a z-component. Thus,

$$T = \frac{|\vec{p}|^2}{2m} = \frac{p_x^2 + p_y^2 + p_z^2}{2m} = \frac{\mathbf{p} \bullet \mathbf{p}}{2m} \tag{2-2}$$

where two forms of notation for a vector quantity have been used, namely $\vec{p}$ and $\mathbf{p}$. If there is a system of N different particles, a subscript serves to distinguish the masses and momenta of the particles, e.g., $m_i$ and $\mathbf{p}_i$ where $i = 1, ..., N$. The kinetic energy for the system is just the sum of the kinetic energies of each of the particles:

$$T = \sum_{i=1}^{N} \frac{\mathbf{p}_i \cdot \mathbf{p}_i}{2m_i} \tag{2-3}$$

Written in this way, the kinetic energy appears as an explicit function of the momentum components of each of the particles; that is, $T = T(\mathbf{p}_1, \mathbf{p}_2, \ldots, \mathbf{p}_N)$.

The potential energy, V, of a conservative system of particles must depend only on the position coordinates of the particles. For example, the potential energy for two electrically charged particles, one with charge $q_1$ and the other with charge $q_2$, is $q_1q_2/r$, where r is the separation distance between the particles. If a Cartesian coordinate system is used to specify particle positions, then the first particle's position may be given by $(x_1, y_1, z_1)$ and the second one's position by $(x_2, y_2, z_2)$. Thus, the potential energy in this case is a function of these six coordinates:

$$V = V(x_1, y_1, z_1, x_2, y_2, z_2) = \frac{q_1 q_2}{\sqrt{(x_2 - x_1)^2 + (y_2 - y_1)^2 + (z_2 - z_1)^2}}$$

It is not required that position coordinates be Cartesian coordinates, though this is often the most convenient. In some important examples, spherical polar coordinates will be used. It is a good idea, then, to carry out further mechanical developments without specifying the type of coordinate system, merely requiring that it be sufficient to specify the positions of all the particles in the system. This is usually termed a *generalized coordinate system*. It is typical notation to use q for a generalized coordinate, or $q_i$ for one from a set of generalized coordinates. The number of degrees of freedom is independent of the choice of coordinate system, and so for a single particle the three possible directions could be denoted by the coordinates $q_1$, $q_2$, and $q_3$. Each generalized position coordinate defines a direction in which there may be a component of momentum. (For every q there is a p.)

The *Hamiltonian*, H, is defined to be simply the sum of T and V, and so it is a function of position and momentum coordinates. For the systems being considered, the Hamiltonian is the total energy of the system.

$$H = T + V \tag{2-4}$$

For nonconservative systems, the potential will have an explicit dependence on time, and so the Hamiltonian will have to be a function of time as well.

Hamilton determined that for a generalized coordinate system, the equations of motion can be found from H:

$$\frac{\partial H}{\partial q_i} = -\dot{p}_i \tag{2-5}$$

$$\frac{\partial H}{\partial p_i} = \dot{q}_i \tag{2-6}$$

where a dot over a character signifies the first derivative of the quantity with respect to time. For each direction, the partial derivative of the Hamiltonian with respect to a position coordinate is equal to the negative of the time derivative of the corresponding (or *conjugate*) momentum coordinate. And the partial derivative of the Hamiltonian with respect to a momentum coordinate equals the time derivative of the conjugate position coordinate. Simultaneous solution of these differential equations yields the description of the mechanics of the system.

## 2.2 THE CLASSICAL HARMONIC OSCILLATOR

A first example of how to use Hamilton's classical equations of motion is in the application to the mechanics of the *harmonic oscillator*. The harmonic oscillator is a special model problem that will be used again and again. It is the problem of a mass able to move in one direction and connected by a spring to an infinitely heavy wall, as shown in Fig. 2.1. The spring is special because it is harmonic, which means that the restoring force is linear in the coordinate that gives the displacement from the equilibrium. An equivalent statement is that the potential energy for stretching or compressing the spring varies quadratically with the distance of displacement from the equilibrium length. If x is chosen to be the position coordinate of the mass, and $x_0$ is the equilibrium length of the

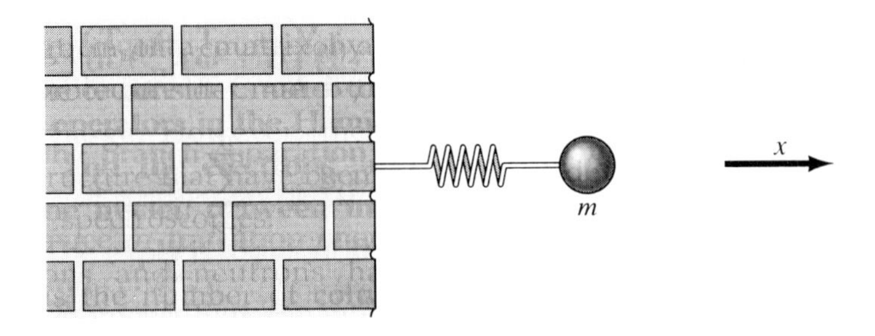


**FIGURE 2.1** The harmonic oscillator is a particle attached by a harmonic spring to an unmovable wall.

spring, then the harmonic potential experienced by the mass in Fig. 2.1 has the following mathematical form.

$$V(x) = \frac{1}{2} k \left(x - x_o\right)^2 \tag{2-7}$$

The constant, k, in this expression is called the *spring constant* or the *force constant*.

Since there is but one degree of freedom for the harmonic oscillator, the kinetic energy is simply $p^2/2m$. Thus, the Hamiltonian for this system is

$$H(x,p) = \frac{p^2}{2m} + \frac{1}{2} k \left(x - x_o\right)^2$$

Differentiating this Hamiltonian function with respect to x and using Eq. (2-5) yields

$$k\,(x - x_o) = -\dot{p}$$

Then, differentiating H with respect to p and using Eq. (2-6) yields

$$\frac{p}{m} = \dot{x}$$

The second of these two equations of motion is the definition of the momentum in Newtonian mechanics; that is, multiplying through by m gives p as equal to the mass times the velocity, $p = m\dot{x}$. Solving the equations of motion is done by any standard means of solving differential equations. In this particular example, the first step is to take the time derivative of the second equation (after multiplying through by m):

$$\dot{p} = m\ddot{x}$$

This relates the time derivative of the momentum to the acceleration or the second time derivative of the position function. But the first equation of motion also tells us that the time derivative of the momentum is $-k(x - x_o)$. Therefore, these must be equal to each other:

$$-k(x - x_o) = m\ddot{x}$$

This expression is the differential equation that determines the position function, x(t).

Hamilton's formulation, applied to the harmonic oscillator problem, has yielded a differential equation that relates the position (a function of time) to the second time derivative of the position. This differential equation has a general solution that may be expressed in either of two ways. They are mathematically equivalent:

$$x(t) = x_o + a\sin(\omega t) + b\cos(\omega t) \tag{2-8a}$$

$$x(t) = x_o + A\,e^{-i\omega t} + B\,e^{i\omega t} \qquad (i \equiv \sqrt{-1}) \tag{2-8b}$$

The constants in these functions are a, b, and $\omega$ in the first, and A, B, and $\omega$ in the second. $x_o$ is the position at which the potential is zero; it is the equilibrium length of the spring. The constant $\omega$ must have units of inverse time, and it is a frequency. The other constants must have units of length because x(t) is the position of the particle. Second differentiation of either Eq. (2-8a) or (2-8b) yields

$$\ddot{x}(t) = -\omega^2\left(x(t) - x_o\right) \tag{2-9}$$

Comparing this with the differential equation derived for the harmonic oscillator demonstrates that the frequency is simply related to the force constant and the particle's mass.

$$\omega = \sqrt{\frac{k}{m}} \tag{2-10}$$

Thus, if the force constant were made larger, i.e., if the spring were stiffer, then Eq. (2-10) says that the frequency would be increased. If the mass were heavier, then the system would be more sluggish, which is to say that the frequency of oscillation would be diminished.

The constants a and b, or A and B, are found from initial conditions. For instance, if at time $t = 0$, it is known that $x(0) = x_0$, then the following is obtained:

$$x(0) = x_0 + a\sin(0) + b\cos(0) = x_0 \quad \Rightarrow \quad b = 0 \qquad (2\text{-}11)$$

(The symbol ⇒ will be used to mean "implies that" or "leads to.") The remaining unknown constant, a, is given by the maximum displacement from equilibrium that the spring reaches. In other words, a is how far the spring had been stretched or compressed initially before it was released to vibrate freely.

At maximum displacement, all the energy of the oscillator is potential energy. This is when the mass is at one of its *turning points*, either $x_0 + a$ or $x_0 - a$. Notice that a is in no way restricted to any particular value. It may be zero, meaning the oscillator is not moving in time. It may be 1.0 or 1.0234 or 120,000.1. This statement just reinforces everyday experience that we can stretch a spring to most *any* length desired and release it in the classical world, at least so long as the spring does not break. The significance of this is that we can adjust the energy of the oscillating spring system in a continuous manner. It will turn out that that only applies to the *classical* harmonic oscillator; this does not happen in the quantum world.

## 2.3 MOTION THROUGH SEVERAL DEGREES OF FREEDOM

Molecular problems are problems of many particles, and so there are many degrees of freedom. To illustrate how Hamilton's formulation is used with more than one degree of freedom, consider the double oscillator problem depicted in Fig. 2.2. This is a system with two degrees of freedom, the positions along the x-axis of the two masses. The subscripts 1 and 2 are

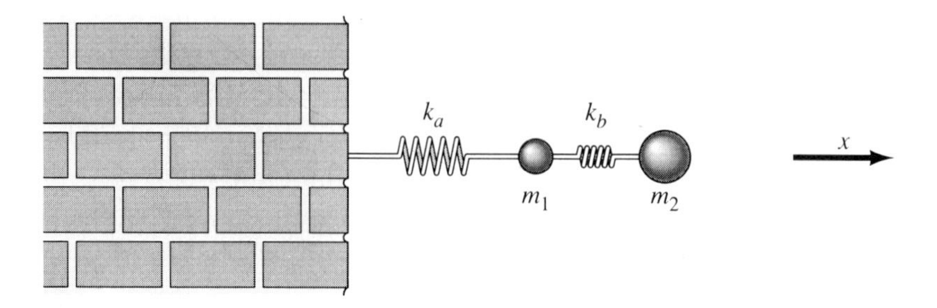


**FIGURE 2.2** A double harmonic oscillator. Both particles 1 and 2 move along the x-axis. Spring a is connected to an unmovable wall and to particle 1. Spring b connects the two particles.

---

used here for the masses, $m_1$ and $m_2$, and for their positions, $x_1$ and $x_2$. The subscripts a and b are used for the two harmonic springs, with force constants $k_a$ and $k_b$ and with equilibrium lengths $x_a$ and $x_b$. The classical Hamiltonian for this system is the sum of the kinetic energies of the two particles and the potential energies from stretching the two springs. We analyze spring b by noting that its length is $x_2 - x_1$. Its displacement from equilibrium is this length less $x_b$, assuming that $x_2 > x_1$ always. The potential energy stored in spring b is this displacement squared, times $k_b/2$. Thus, the Hamiltonian is

$$H(x_1,x_2,p_1,p_2) = \frac{p_1^2}{2m_1} + \frac{p_2^2}{2m_2} + \frac{1}{2}k_a(x_1 - x_a)^2 + \frac{1}{2}k_b(x_2 - x_1 - x_b)^2 \tag{2-12}$$

The equations of motion, by Hamilton's formulation, follow from Eqs. (2-5) and (2-6).

$$p_1 / m_1 = \dot{x}_1 \tag{2-13a}$$

$$k_a(x_1 - x_a) - k_b(x_2 - x_1 - x_b) = -\dot{p}_1 \tag{2-13b}$$

$$p_2 / m_2 = \dot{x}_2 \tag{2-14a}$$

$$k_b(x_2 - x_1 - x_b) = -\dot{p}_2 \tag{2-14b}$$

These are *coupled* differential equations that reflect the fact that the two springs are physically coupled: They are connected at mass 1. Actually solving this type of problem or this set of coupled differential equations is considered at the end of this chapter. For now, the important feature is that systems of several degrees of freedom will, in general, have coupled equations of motion.

There are circumstances where the equations of motion in several degrees of freedom are not coupled. In these cases, the Hamiltonian has a special form that shall be referred to as a *separable* form. Separability arises when a particular Hamiltonian may be written as additive, independent functions, e.g.,

$$H(q_1,q_2,q_3,\ldots,p_1,p_2,p_3,\ldots) = H'(q_1,p_1) + H''(q_2,q_3,\ldots,p_2,p_3,\ldots) \tag{2-15}$$

In this instance, H' is a function only of the first position and momentum coordinates, while H" is a separate function involving only the other coordinates. If we apply Eqs. (2-5) and (2-6) to H in order to find the equations of motion for this system, the equations for $q_1$ and $p_1$ are found to be distinct, or mathematically unrelated to any of the equations that involve any of the other coordinates. In fact, the equations for $q_1$ and $p_1$ could just as well have been obtained by taking H' to be the Hamiltonian for the motion in the $q_1$-direction, while the other equations of motion could have been obtained directly from the H" Hamiltonian. Thus, the additive, independent terms in the Hamiltonian may be separated in order to deduce the equations of motion. Furthermore, since the equations of motion are separable, then the actual physical motions of the H' system and the H" system are unrelated and independent. It is possible that a Hamiltonian may be separable in all variables; that is, it may be that

$$H(q_1,q_2,q_3,\ldots,p_1,p_2,p_3,\ldots) = H_1(q_1,p_1) + H_2(q_2,p_2) + H_3(q_3,p_3) + \ldots$$

Then all the motions are independent.

In the double-oscillator system of Fig. 2.2, the last potential energy term involved both position coordinates together, and so that problem was not separable in $x_1$ and $x_2$. To contrast with that situation, we will next consider a problem that ends up displaying separability. It is the oscillator problem in Fig. 2.3. The Hamiltonian is

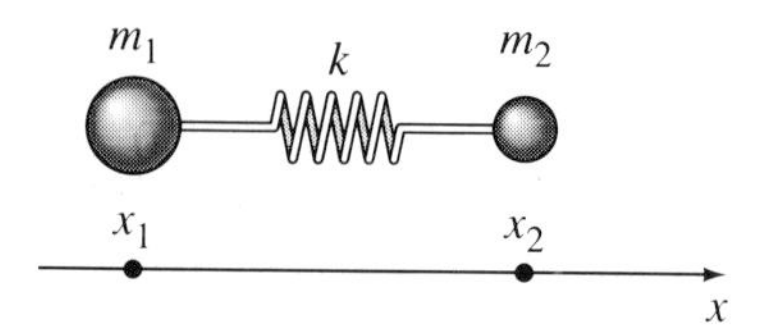


**FIGURE 2.3** An oscillator system where two particles, each constrained to move along the x-axis, are attached by a harmonic spring with force constant k. The positions of the two particles along the x-axis are designated $x_1$ and $x_2$. The equilibrium length of the spring is the constant, $x_o$.

$$H(x_1,x_2,p_1,p_2) = \frac{p_1^2}{2m_1} + \frac{p_2^2}{2m_2} + \frac{1}{2}k(x_2 - x_1 - x_o)^2$$

Because of the potential energy term, this appears to be inseparable. However, it turns out that a certain change of coordinates, or more precisely a *coordinate transformation,* does lead to a separable form. This coordinate transformation comes about by defining two new variables. First, we define r to be a coordinate that gives the displacement from the equilibrium length of the spring, and second, we define s to be the position of the center of mass of the whole system:

$$r \equiv x_2 - x_1 - x_o$$

$$s \equiv \frac{m_1x_1 + m_2x_2}{m_1 + m_2}$$

By inspection, the potential energy now has the simple form $kr^2/2$. That is, in the {r,s} coordinate system, the potential energy is independent of s, the position of the system's mass center.

The kinetic energy of this system can be expressed in terms of momentum coordinates that are conjugate to the position coordinates, r and s. First, Eq. (2-5) is used in the $\{x_1,x_2\}$ system to write

$$p_1 = m_1\dot{x}_1 \qquad p_2 = m_2\dot{x}_2$$

Next, the kinetic energy is expressed in terms of the time derivatives of the position coordinates by using these expressions for the momenta.

$$T = \frac{1}{2}(m_1 \dot{x}_1^2 + m_2 \dot{x}_2^2)$$

Next, the time derivatives of r and s are substituted for the time derivatives of $x_1$ and $x_2$. The relation between velocities in the two coordinate systems comes about from taking the time derivative of the equations that relate r and s with $x_1$ and $x_2$:

$$\dot{r} = \dot{x}_2 - \dot{x}_1$$

$$\dot{s} = (m_1 \dot{x}_1 + m_2 \dot{x}_2) / (m_1 + m_2)$$

By rearrangement, we may obtain from these two equations that

$$\dot{x}_1 = \dot{s} - \frac{m_2}{(m_1 + m_2)} \dot{r}$$

$$\dot{x}_2 = \dot{s} + \frac{m_1}{(m_1 + m_2)} \dot{r}$$

Using these expressions for $\dot{x}_1$ and $\dot{x}_2$ in the equation for T yields

$$T = \frac{1}{2}\left[ (m_1 + m_2)\dot{s}^2 + \frac{m_1 m_2}{(m_1 + m_2)} \dot{r}^2 \right]$$

At this point, M will be used to designate the total mass of the system, i.e., $M = m_1 + m_2$, and $\mu$ will be used for the quantity $m_1 m_2 / (m_1 + m_2)$. $\mu$ is called the *reduced mass* of a two-body system. And so,

$$T = \frac{1}{2}\left( \mu \dot{r}^2 + M \dot{s}^2 \right)$$

Notice that this is similar in form to the kinetic energy expressed in the original coordinates $x_1$ and $x_2$. The similarity suggests a definition of the conjugate momenta in the {r,s} coordinate system:

$$p_r = \mu \dot{r}$$

$$p_s = M \dot{s}$$

In fact, it is Hamilton's equations of motion [Eq. (2-5)] that ensures that these are correct definitions.

The Hamiltonian for the oscillator of Fig. 2.3 can now be written fully transformed to the {r,s} coordinate system.

$$H(r, s, p_r, p_s) = \frac{p_s^2}{2M} + \frac{p_r^2}{2\mu} + \frac{1}{2} k r^2$$

This is recognized to be a separable Hamiltonian because the first term is independent of the remaining two terms. The first term is interpreted as the kinetic energy of translation of the whole system. The second term is the kinetic energy for the *internal* motion of vibration, while the last term is the potential energy for vibration. Separability means that the translational motion of the whole system is independent and unrelated to the internal motion.

As discussed above, separability of a Hamiltonian means that the equations of motion could just as well have been developed from the independent terms if they had been treated as individual Hamiltonians. For the example of Fig. 2.3, this says that the equations of motion obtained from $H(r,s,p_r,p_s)$ will be the same as those obtained for the separate problems of internal motion, where the Hamiltonian is

$$H'(r, p_r) = \frac{p_r^2}{2\mu} + \frac{1}{2} k r^2$$

and of translational motion, for which the Hamiltonian is

$$H''(s, p_s) = \frac{p_s^2}{2M}$$

In other words, the four equations of motion we would obtain for the entire system using H turn out to be the same as the two equations of motion obtained from the H' function and the two equations of motion obtained from the H" function. This comes about because the vibration of the system, which is motion in the r-direction, does not affect and is not affected by the translation of the system, which is motion in the s-direction. So, H' alone is sufficient if only internal motion mechanics are of interest; translational motion of the system may be ignored because it is separable.

Comparison of the internal vibrational Hamiltonian for the two-body oscillator of Fig. 2.3 with the Hamiltonian for the one-body oscillator of Fig. 2.1 shows that the differences are only in the "names" of the masses, m versus $\mu$, and the names of the displacement coordinates, x versus r. On this basis, we may say that vibration of the two-body system is mechanically equivalent to vibration of a single body of mass $\mu$ attached to an unmovable wall. *Mechanical equivalence* means two systems have equations of motion (or Hamiltonians) of the same form.

The examples so far have been with particles that have been restricted to move only in a straight line, along the x-axis. A more general analysis is necessary, of course, since atoms and molecules do not have to experience such constraints. A free atom may move in the x-direction or the y-direction or the z-direction. Should any other type of coordinate system be used to give the location of the atom, there will still be three and only three independent coordinates. For a system composed of N atoms, or N particles, whatever the type, there will be a total of 3N independent coordinates required to specify the positions of all N particles. There will be 3N *degrees of freedom* for the system.

We have seen that the translational motion of a system of two particles along the x-axis is separable from the vibrational motion. In a three-dimensional picture of the system, translational motion is also separable, but the coordinate transformation is different. In three Cartesian dimensions, the positions of the two particles may be specified as $(x_1,y_1,z_1)$ and $(x_2,y_2,z_2)$. The center of mass of the system is also given by three coordinate values (X,Y,Z), and they are

$$X = \frac{1}{M}(m_1 x_1 + m_2 x_2)$$

$$Y = \frac{1}{M}(m_1 y_1 + m_2 y_2) \qquad (2\text{-}16)$$

$$Z = \frac{1}{M}(m_1 z_1 + m_2 z_2)$$

The separation distance between the two particles, r, is the following.

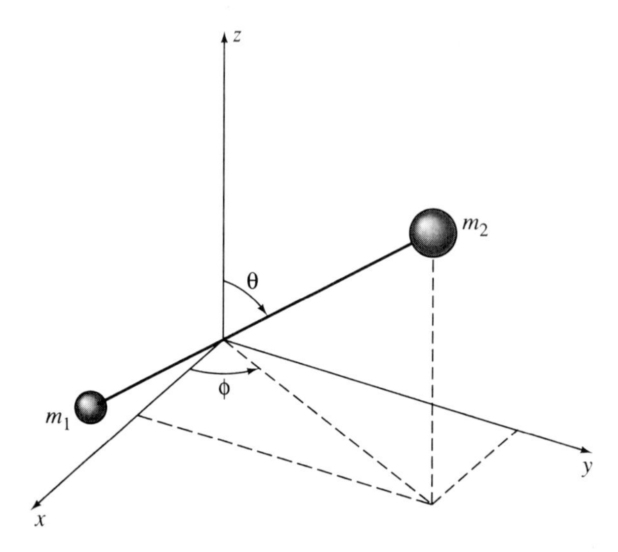


**FIGURE 2.4** Spherical polar coordinates for a two-particle system. r is the distance between $m_1$ and $m_2$. The center of mass is positioned at the origin of the (x,y,z) coordinate system. θ is the angle between the z-axis and the line connecting the two masses. θ ranges from 0 to π. φ measures the rotation *about* the z-axis of the line connecting the masses. It ranges from 0 to 2π.

---

$$r = \sqrt{(x_2 - x_1)^2 + (y_2 - y_1)^2 + (z_2 - z_1)^2} \tag{2-17}$$

Two other new coordinates, θ and φ, complete the transformation. They give the orientation of the two-particle system about the center of mass. As depicted in Fig. 2.4, r, θ, and φ are *spherical polar coordinates.*

$$\begin{aligned} x_2 - x_1 &= r \sin\theta \cos\phi \\ y_2 - y_1 &= r \sin\theta \sin\phi \\ z_2 - z_1 &= r \cos\theta \end{aligned} \tag{2-18}$$

The overall coordinate transformation is $\{x_1, y_1, z_1, x_2, y_2, z_2\} \rightarrow \{X, Y, Z, r, \theta, \phi\}$. Notice that the number of independent coordinates, six, remains the same.

The transformation to spherical polar coordinates makes the potential energy dependent only on the variable, r, rather than on all six coordinates. The kinetic energy has a piece associated with translational motion of the whole system,

$$T^{\text{translation}} = \frac{1}{2M}(p_X^2 + p_Y^2 + p_Z^2)$$

It also has a piece associated with the vibrational motion and a piece associated with rotation of the two bodies about the center of mass. Thus, the Hamiltonian will consist of independent terms associated with motion in the X-direction (i.e., translation of the system along the x-axis), in the Y-direction, and in the Z-direction, and with vibrational and rotational motion. This describes the separability of the problem. There are three translational degrees of freedom that are separable from the rest.

Had we analyzed a system of N particles with $N > 2$, there would still be only three translational degrees of freedom for the system as a whole. And this motion would be separable from the other motions. Thus, there would be $3N - 3$ degrees of freedom for other-than-translational motion. In the case of two point-mass particles, there were two rotational degrees of freedom corresponding to the two angles $\theta$ and $\phi$. This will be the case for any linear system, and so there will be $(3N - 3) - 2 = 3N - 5$ vibrational degrees of freedom. For nonlinear collections of particles, there are three independent rotations about the mass center since rotations about the x-axis, about the y-axis, and about the z-axis are all different. Thus, for nonlinear systems, $3N - 6$ is the number of degrees of freedom remaining for vibrational motion.

It may not seem clear why there should be a different number of rotational degrees of freedom for linear and for nonlinear systems. The subtle difference has to do with the fact that we are considering particles to be point-masses, entities without volume or size. If some number of these masses were arranged along the x-axis, we could certainly visualize turning the system about the y-axis or turning it about the z-axis. On the other hand, turning it about the x-axis is to do nothing. It is an operation or motion that does not exist. Only if one or more particles were off-axis would such motion be defined, but then that would be a nonlinear arrangement.

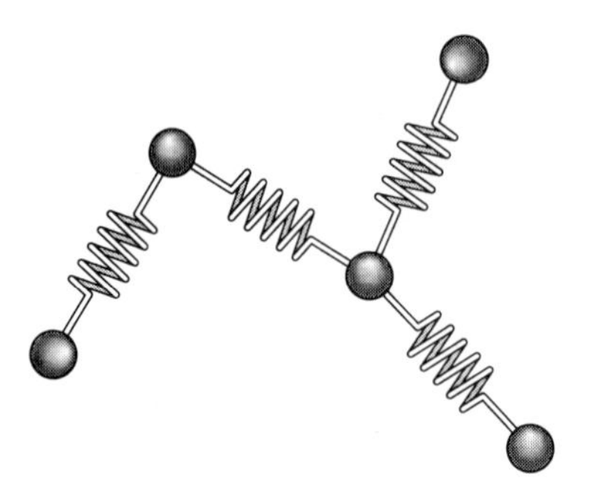

**FIGURE 2.5** A vibrating system of particles connected by springs. The potential energy of this system is a sum of the potential energy stored in each spring. Since the energy of a given spring has a quadratic dependence on the position coordinates of the masses that it connects, then in this system, the motions of the particles are surely coupled by the springs.

## 2.4 HARMONIC VIBRATION OF MANY PARTICLES

The oscillator problem of Fig. 2.2 yielded a nonseparable Hamiltonian because the two springs were connected to one of the masses and this coupled the stretching and contracting of the springs. The corresponding mathematical feature is a cross-term in the potential energy expression, a term that involves both of the position coordinates, $x_1$ and $x_2$. This problem is representative of the complexity of the problems of molecular vibration since we could imagine the chemical bonds of a molecule behaving much like springs connecting atoms. The vibrational potentials of some molecule considered in this way (Fig. 2.5) would have plenty of cross-terms of the atomic displacement coordinates. Pulling a bit on one atom would obviously affect the other atoms. All the motions are coupled through the network of springs, and were the Hamiltonian to be written in terms of atomic position coordinates, it would not be separable. However, the oscillator problem of Fig. 2.3 also had a cross-term involving the particle position coordinates, and yet that problem was finally separated into internal vibration and the translation of the whole system. A coordinate transformation was the key to obtaining that Hamiltonian in separable form. So, we may ask, could a coordinate transformation lead to separability for the general type of problem in Fig. 2.5? In fact, there is a specific coordinate transformation that can accomplish just that. However, the process of making that transformation assumes that all the springs are harmonic, and so when we later apply this process to molecules, it will be as an approximation of the behavior of chemical bonds. The potentials for chemical bonds are always at least somewhat anharmonic, meaning that there is a cubic, quartic, or higher order dependence of the potential energy on the extent of extension or compression of the "spring" (bond).

If separability of the harmonic vibrations is achieved, the Hamiltonian in the final coordinate system must have the following general form.

$$H = \sum_i \left( \frac{1}{2} \alpha_i q_i^2 + \frac{p_i^2}{2\beta_i} \right) \qquad (2\text{-}19)$$

In other words, complete separability means that each degree of freedom has an independent term in the Hamiltonian. Just the form of this Hamiltonian tells much about the vibrations of the system, and this can be seen by thinking about a system that is mechanically equivalent. That system would consist of a set of different harmonic oscillators, each of the type shown in Fig. 2.1 and none connected to another. For the $i^{th}$ oscillator of this set, there would be a displacement coordinate, $q_i$, a spring constant which could be called $\alpha_i$, and the particle's mass which could be called $\beta_i$. One could initiate vibration by stretching and releasing any one of these oscillators, and there would be no effect on the others. By mechanical equivalence, the same is true for the coupled harmonic oscillator problem (Fig. 2.5); however, the $q_i$-coordinates are more complicated because they turn out to be linear combinations of the many different particle displacement coordinates. A given $q_i$ could correspond to a displacement of all the particles at once.

The coordinates $\{q_i\}$ for which the vibrational Hamiltonian takes on the separable form of Eq. (2-19) are referred to as *normal coordinates*. Motion that follows along the direction of these coordinates is referred to as *normal mode* vibration. For each normal mode, there is a vibrational frequency and, by mechanical equivalence arguments, its value is

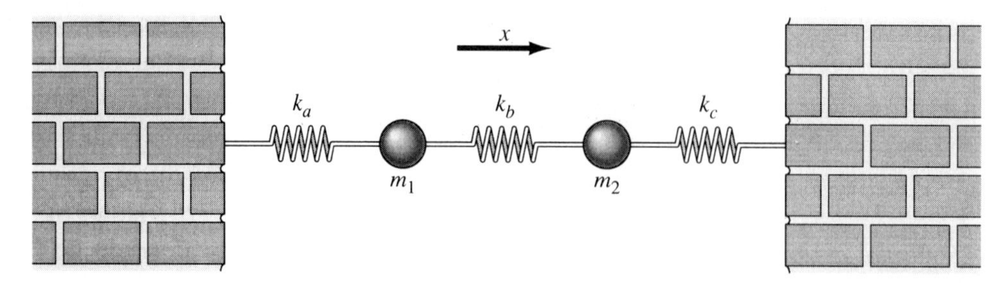


**FIGURE 2.6** A vibrating system of two particles constrained to move in the x-direction and connected by springs to each other and to two unmovable walls.

$$\omega_i = \sqrt{\frac{\alpha_i}{\beta_i}} \tag{2-20}$$

The important quantities $\alpha_i$ and $\beta_i$ depend on the masses of the particles and the force constants of the springs, and they do so in a way that will have to be worked out through coordinate transformations.

Fig. 2.6 shows another oscillator problem that involves two particles. If the particles are of the same mass and if the three springs happen to have the same force constant, the two normal coordinates have a simple qualitative form. One corresponds to the two particles moving together; at the same time, both are moving to the left or else both are moving to the right. The other normal coordinate corresponds to the two particles moving oppositely; if one is moving to the right, the other is moving to the left. These two normal modes of vibration have different frequencies of oscillation. This fact is physically reasonable because in the parallel motion the length of the central spring does not change. Only two springs are stretched or contracted. In the opposed motion, when the two outer springs are squeezed, the inner one is stretched. The system is "stiffer" with respect to the latter type of motion, and it is correct to expect the frequency of that vibrational mode to be greater than the frequency of the parallel motion mode.

With the example of Fig. 2.6 in mind, we can understand several features of a normal mode description. If the system is at rest at its equilibrium structure initially, and then the particles are displaced by an amount that is proportional to one normal coordinate, releasing the particles will cause the system to vibrate only in that mode. In the Fig. 2.6

example, with equal masses and like springs, displacement from equilibrium of one particle to the left and the other to the right by the same amount is displacement proportional to one of the normal coordinates. Displacement of both particles to the right (or left) by the same amount would be a displacement along the other normal coordinate.

When a system vibrates in a pure mode, all the particles in the system will reach their point of maximum displacement at the same instant. They will all pass through their equilibrium positions at the same instant. Simply, they will all vibrate with the same frequency and phase. In carbon dioxide, for instance, the pure symmetric stretching vibrational mode is one that equivalently stretches both carbon-oxygen bonds simultaneously. In this vibrational mode, the carbon atom remains at the midpoint of the line between the oxygen atoms as they oscillate back and forth with the same frequency.

A system such as that in Fig. 2.6 will normally vibrate in a more complicated manner than simple normal mode motion even if the springs are perfectly harmonic. However, any such motion can be represented as a superposition of the different normal mode motions.

## Exercises

1. Set up the classical Hamiltonian for a particle of mass m that may move in the three directions, x, y, and z, and that experiences a gravitational potential of the form $V(x,y,z) = mgz$. Then, using Eqs. (2-5) and (2-6), write the equations of motion for this particle.
2. Write the explicit form of the Hamiltonian for a system of two charged particles connected by a massless harmonic spring with force constant k. Use $q_1$ and $q_2$ for the charges of the particles and $m_1$ and $m_2$ for their masses.
3. Show that a potential of the general form $V(x) = a + bx + cx^2$ is the same as that of a harmonic oscillator because it can be written as $V(x) = V_o + \frac{1}{2} k (x - x_o)^2$. (Find k, $V_o$, and $x_o$ in terms of a, b, and c.)
4. Verify Eq. (2-9) with both Eqs. (2-8a) and (2-8b).

5. From the exponential relation that for any real number $\alpha$, $e^{i\alpha} = \cos(\alpha) + i\sin(\alpha)$, find the constants A and B in Eq. (2-8b) so that this x(t) is the same as the x(t) in Eq. (2-8a) for the special case when b = 0 and a = 1. Next, generalize your result for the constant a left unspecified. This amounts to writing A and B as a function of a.
6. What force constant for a harmonic spring would give rise to a vibrational frequency, $\omega$, of 10 $\text{sec}^{-1}$ when a mass of 1 kg is attached? What would be the frequency if such a spring were attached to a particle with the mass of (i) an electron, (ii) a hydrogen atom, (iii) the moon?
7. Write the potential energy of the following system of point-mass particles and harmonic springs as a function of the position coordinates of the particles. Assume that they are constrained to move only in the xy plane.

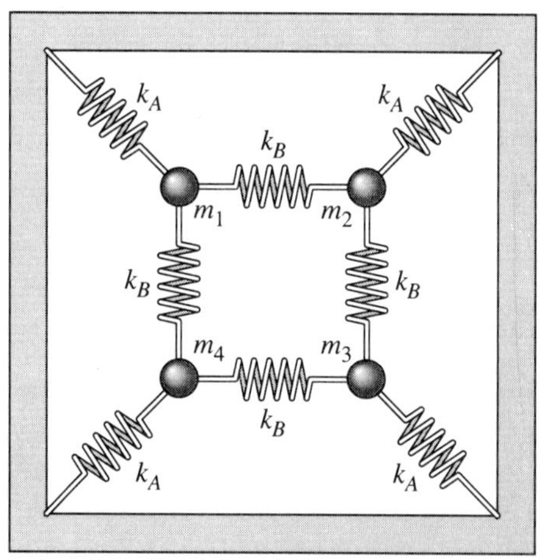


Let $l$ be the length of each side of the square, let $s_a$ be the equilibrium length of the springs with force constant $k_a$, and let $s_b$ be the equilibrium length of the springs with force constant $k_b$.

8. Develop Hamilton's equations of motion for the system in Prob. 7.
9. Set up the Hamiltonian for a particle of mass m that is connected to a harmonic spring with force constant k. The other end of the spring is connected to a pivot that allows the spring-and-mass assembly to turn about freely in any direction. The location of the pivot is fixed in space and does not move. Show that this system is mechanically equivalent in vibrational and rotational motion to a system of two

particles connected by the same spring, assuming the particles may move in all three dimensions.

*10.* In what qualitative manner would a symmetric stretch mode of the system in Prob. 7 differ in having $m_1 = m_2 = m_3 = m_4$ and in having $m_1 = 2m_2 = m_3 = 2m_4$?

*11.* What would be the qualitative forms of the normal modes of vibration of the oscillator in Fig. 2.6 if there were not two but three equally massive particles connected one to another and to the walls by harmonic springs with the same force constants? And if there were four particles?

*12.* If the two particles in Fig. 2.6 were considered more realistically, there would be a y- and a z-degree of freedom for each particle to move. This would mean four more vibrational degrees of freedom, and they would be associated with bending modes of the system. Qualitatively, what would they look like? Should any of them have the same normal mode frequency?

## Bibliography

1. K. R. Symon, *Mechanics* (Addison-Wesley, Reading, Massachusetts, 1971). This is an intermediate level text in classical mechanics, and its coverage goes beyond our immediate interests. The sections on Hamilton's formulation of classical mechanics and the sections on normal modes of vibration, though, may be quite helpful.

# Chapter 3

# Quantum Mechanics - I

*The classical mechanics of the last chapter proves deficient for describing systems of very small particles such as atoms and electrons. Instead, a very different physical picture and mathematical apparatus are needed, and these are embodied in quantum mechanics. This chapter explores some of basic elements of quantum mechanics and also considers the idealized problem of a harmonic oscillator.*

## 3.1 QUANTUM PHENOMENA

The first of what are now called quantum phenomena were noticed or detected around the beginning of the twentieth century. Numerous experiments could not be explained by the electromagnetic and mechanical laws known or accepted at the time. For example, it was found that a beam of monoenergetic electrons could be diffracted in the same way as a monochromatic beam of light impinging on two slits. The diffraction of light was known to be a manifestation of its wave character, and so to explain the diffraction of electrons, an abstract type of wave outside the paradigm of classical mechanics, which holds that particles are exactly what we sense and observe in our everyday world: They are distinct objects that exist at a specific place at any instant of time, unlike electromagnetic radiation, which is distributed in space and time.

It was no doubt unsatisfying to many scientists during the early part of the century that basic physical laws worked out over almost two centuries did not hold for the tiny world of electrons and atoms. Today, we see the classical picture as a special case of broader theories of matter and radiation. It took studies of the tiny-particle world and the introduction of several new concepts to develop the more encompassing theories. This was because key features of the quantum world are manifested in ways not directly perceptible in our everyday world.*

Particle diffraction was an important observation. The particles are free particles meaning they are not subject to a potential. Electrons, for instance, would first have been stripped away from some atoms or molecules. Observing diffraction requires that all the particles in the sample move at the same velocity or momentum, and this can be accomplished by the manner in which the particle beam is formed. The quantitative result of the diffraction experiment is a wavelength, $\lambda$. Measured for different particle momenta, p, the wavelength is found to vary inversely with p. This is called the *de Broglie relation*:

$$p = \frac{h}{\lambda} \tag{3-1}$$

$\lambda$ is the de Broglie wavelength. The proportionality constant of this relation, h, turns out to be a fundamental constant of nature, *Planck's constant*. The value of h is $6.6256 \times 10^{-27}$ erg-sec.

Let us find the de Broglie wavelength for an object from our day-to-day world, a rider on a motorcycle traveling at 36 km/hr. If the mass of the rider and cycle is 200 kg, then the momentum of this "particle" is

* A question that often comes to mind upon first learning of the quantum mechanical revolution is whether a new theory could someday replace quantum mechanics. In this day and age, this may be more a matter for the philosophy of science and how science accepts or rejects new theories, for the successes of quantum ideas appear endless and have stood the test of time. There is good reason for confidence that any major scientific revolution to come will not replace quantum theory but will build upon or encompass it.

$$(200 \text{ kg}) \times (36 \text{ km/hr}) = (2 \times 10^5 \text{ g}) \times (10^3 \text{ cm/sec})$$
$$= 2 \times 10^8 \text{ g-cm/sec}$$

Dividing this into Planck's constant yields the wavelength.

$$\lambda = \frac{6.63 \times 10^{-27} \text{ g-cm}^2/\text{sec}}{2 \times 10^8 \text{ g-cm/sec}} = 3.3 \times 10^{-35} \text{ cm}$$

This is an incredibly tiny wavelength, and it is not surprising that effects associated with something this size are not within the perception of the motorcycle rider. However, the de Broglie wavelength of a free electron moving at the same velocity as the rider and cycle is much longer:

$$\lambda = \frac{6.63 \times 10^{-27} \text{ g-cm}^2/\text{sec}}{(9.1 \times 10^{-28} \text{ g})\,(10^3 \text{ cm/sec})} = 7.3 \times 10^{-3} \text{ cm}$$

The wavelength of the electron, because of its small mass, is significant relative to the dimension of its atomic world, whereas the wavelength of the rider and cycle is negligible relative to the dimensions of their world. One may consider it a valid approximation to treat the mechanics of the rider and cycle classically, whereas the wave character of the free electron is not so ignorable.

Other observations made at the beginning of the twentieth century not only showed unexpected phenomena but also seemed to involve the new constant, h, in many ways. The photoelectric effect, which is the emission of electrons from a metal surface upon irradiation with light, was explained by Einstein with a nonclassical picture of electromagnetic radiation. He argued that the energy delivered by radiation comes in discrete amounts, called quanta, and that the energy is proportional to the frequency of the radiation, $\nu$:

$$E^{\text{radiation}} = h\nu \tag{3-2}$$

Notice that the proportionality constant is Planck's constant. As more problems were explained by the unusual quantum and wave notions, it became certain that classical physics was not complete and could not be applied at the subatomic level. The laws of quantum mechanics that then emerged were worked out by clever and sometimes indirect lines of thought.

## 3.2 WAVE CHARACTER

If it is necessary for a quantum-level description that a free particle have an associated wavelength, then there should be some simple oscillating function associated with the particle's description. This would be a sine or cosine function (or equivalently, a complex exponential), e.g.,

$$A(x) = A_o \cos\left(\frac{2\pi x}{\lambda}\right) \tag{3-3}$$

The constant, $A_o$, is the maximum amplitude of A(x). It is a feature of this type of function that the wavelength may be extracted from the function as an *eigenvalue*. Notice that when A(x) is differentiated twice, the result is a constant, the eigenvalue C, times A(x).

$$\frac{d^2}{dx^2} A(x) = -\left(\frac{2\pi}{\lambda}\right)^2 A_o \cos\left(\frac{2\pi x}{\lambda}\right) = C\,A(x)$$

In such a situation this function is called an *eigenfunction* of the operation of second differentiation. A translation of the German word *eigen* is "of one's own," or loosely, "same." It is one's own function, or the same function, that is produced upon the operation of second differentiation. The constant that shows up multiplying the original function is the eigenvalue. In contrast, A(x) is not an eigenfunction of the operation d/dx since first order differentiation of the cosine function produces a different function, the sine function.

At this point we might believe that somehow the function A(x) or something similar is going to be the basis for a description of the mechanics of a moving free particle. However, we do not have an interpretation of A(x); its meaning is not apparent. One thing, though, is that A(x) has a wavelength embedded in it, and that wavelength shows up in a particular eigenvalue of the function. It was deduced by Schrödinger, Heisenberg, and others that eigenvalues have special significance, and in particular that they give the measurable information about a system, such as the de Broglie wavelength of a free particle. The eigenfunctions themselves came to be known as *wavefunctions* partly because it was the apparent wave character of particles that demanded a new mechanics.

If there are particular operations, such as second differentiation, for which all proper wavefunctions must be eigenfunctions, then there exists a general basis for finding wavefunctions for any system. It is to solve an eigenequation – rather than guess at something like A(x). This focuses attention on the operations that may enter the eigenequation, and so, specific mechanical elements come to be associated with specific mathematical operations. One of these can be demonstrated using the de Broglie relation and the free particle example. If the operation $(d/dx)^2$ is scaled (multiplied) by the constant* $-(h/2\pi)^2$, then by Eq. (3-1) the resulting operator yields the square of the momentum when applied to A(x):

$$-\left(\frac{h}{2\pi}\right)^2 \frac{d^2}{dx^2} A(x) = \frac{h^2}{\lambda^2} A(x)$$

This suggests an operation that in quantum mechanics corresponds to the square of the dynamical variable of momentum. We might anticipate that this is useful, since the classical energy, at least, is a function of momentum, and energy is of key importance in a particle world where energy is quantized.

These several notions about the quantum world form part of a theoretical basis that is usually referred to as the set of *quantum mechanical postulates*. These postulates are ideas or statements that in themselves may or may not be true. However, if predictions that follow naturally from these ideas are proven true, then the ideas are accepted as true even if there is no direct proof.

Postulate I says that *for every quantum mechanical system, there exists a wavefunction that contains a full mechanical description of the system*. It is a function of position coordinates and possibly time. This is quite different from classical mechanics where the full mechanical description would be a function that gives the position coordinates for any instant in time, e.g., $\vec{x}(t)$.

Postulate II explains, in part, that we may glean a picture of a quantum mechanical system from its wavefunction. *For every dynamical variable, there is an associated mathematical operation. Furthermore, if*

* Planck's constant, h, often shows up in equations divided by $2\pi$. Thus, a special symbol, $\hbar$, is used to replace the ratio $h/2\pi$. $\hbar$ (h-bar) is also called Planck's constant, and so it is always important to indicate whether h or $\hbar$ is meant.

*there is an eigenvalue for such an operation and a particular wavefunction, then that eigenvalue is the result that would be obtained from measuring that dynamical variable for the particular system.* An example, so far, is that the operation associated with the square of the momentum has an eigenvalue of $(h/\lambda)^2$ for the free particle wavefunction, A(x).

The dynamical variables in the classical Hamiltonian are position and momentum coordinates. As long as wavefunctions are developed* as functions of position coordinates, as was the case for A(x), the mathematical operation associated with a position coordinate is just multiplication by the coordinate. The operation associated with the momentum variable $p_i$ that is conjugate to the position variable $q_i$ is

$$-i\hbar \frac{\partial}{\partial q_i}$$

where $i \equiv \sqrt{-1}$. Notice that the square of this operation (i.e., applying it twice) is the same as that used above for the square of the momentum. The various mathematical operations are best referred to as *operators*, and operators may be distinguished from variables by a "hat" over the character. Thus, the statement $\hat{x} = x$ means that the operator associated with x is the same as the variable x, while $\hat{p} = -i\hbar\, \partial/\partial x$ means that the operator associated with the momentum is $-i\hbar$ times the operation of differentiation with respect to x. Operators can be combined and applied repeatedly to form new operators. Thus, if the position and momentum operators are known, an operator can be constructed that is associated with the classical Hamiltonian. It is called the Hamiltonian operator, or just the Hamiltonian if the quantum mechanical context is clear.

The Schrödinger equation, which can be taken as Postulate III, is that *the wavefunction of a system must be an eigenfunction of the Hamiltonian operator*. The eigenvalue is the energy of the system, since by association, energy is the quantity given by the classical Hamiltonian. Energy is measurable or observable, and so this statement is similar to the second

* Quantum mechanics may also be developed so that the functions describing the system are functions of momentum coordinates, not position coordinates. This is termed a momentum representation, and in this representation, position and momentum operators take on a different form. The picture of a quantum system, though, is equivalent, and the choice between representations is largely one of convenience.

postulate that tells what to make of eigenvalues of wavefunctions. But the third postulate is a stronger statement than the second postulate in that it *requires* the wavefunction to be an eigenfunction of one specific operator, the Hamiltonian. The most common symbol used for an unspecified quantum mechanical wavefunction is $\psi$, and so the Schrödinger equation is written as

$$\hat{H}\psi = E\psi \tag{3-4}$$

E is the energy eigenvalue. Solution of the Schrödinger equation is central to all quantum mechanical problems.

## 3.3 OPERATORS

Because of the important mathematical role of operators in quantum mechanics, it is useful to understand the algebra of operators. As a general definition, we may say that an operator is a mathematical device that relates two functions. An operator equation is any equation that includes an operator. Thus, $g(x) = 3f(x)$ may be regarded as an operator equation, since the functions $g(x)$ and $f(x)$ are related by the operation of multiplication by 3. The equation could be given more formally as $g(x) = \hat{O}\, f(x)$ where $\hat{O} = 3$.

Though operators may very well be as simple as multiplication by a constant such as 3, they may be multiplication by a variable, such as the operator $\hat{x}$. Or they may involve differentiation with respect to one or more variables. Anything that can relate one function to another function can be labelled an operator.

Two operators may be combined by addition or by multiplication to create another operator. The process of operator addition and the process of operator multiplication, though, require definition.

$\hat{c} = \hat{a} + \hat{b}$ means the result of operating with $\hat{c}$ on an arbitrary function, f, yields the same result as operating with $\hat{a}$ on f and adding that to the result of operating with $\hat{b}$ on f.

$\hat{d} = \hat{a}\,\hat{b}$ means that the result of operating with $\hat{d}$ on an arbitrary function, f, yields the same result as operating on f with $\hat{b}$ and then operating on that result with $\hat{a}$.

With these definitions, operators can be added and multiplied, and so algebraic expressions may be written involving just operators.

Multiplication of operators is successive application of two or more operators. However, this is not the same as multiplication of numbers; the order of the operators being multiplied may make a difference. Consider the example where one operator is multiplication by the variable x, $\hat{a} = x$, and the other operator is differentiation with respect to x, $\hat{b} = d/dx$. To invoke the definition of operator multiplication, let f(x) be an arbitrary function; that is, let it stand for any function whatsoever that we may choose. The product of the operator multiplication of $\hat{a}$ times $\hat{b}$ is found from applying $\hat{b}$ and then $\hat{a}$ to the arbitrary function. But this result is seen to be different from multiplication of $\hat{b}$ times $\hat{a}$.

$$\hat{a}\,\hat{b}\,f(x) = x\,\frac{df(x)}{dx} \qquad \Rightarrow \qquad \hat{d} = x\frac{d}{dx}$$

$$\hat{b}\,\hat{a}\,f(x) = \frac{d}{dx}\big(x\,f(x)\big) = x\,\frac{df(x)}{dx} + f(x) \qquad \Rightarrow \qquad \hat{d} = \left(x\frac{d}{dx} + 1\right)$$

This illustrates that operator multiplication is not commutative. Multiplication of simple numbers, of course, is commutative; 2 times 3 is the same as 3 times 2.

A *commutator* is a special operator constructed from two operators as the difference between the two ways they may be multiplied. The commutator is designated by placing the two operators being multiplied inside square brackets. The definition is

$$[\hat{A},\hat{B}] \equiv \hat{A}\,\hat{B} - \hat{B}\,\hat{A} \tag{3-5}$$

If the two operators do commute, which means that the same result is achieved from applying them in either order, then the commutator is equal to zero. On the basis of the multiplication results for the $\hat{a}$ and $\hat{b}$ operators just considered, their commutator is: $[\hat{a},\hat{b}] = -1$.

A special type of operator equation is the eigenvalue equation, which may be expressed generally as

$$\hat{O} f = c f \tag{3-6}$$

This equation says that for some particular function f and some operator $\hat{O}$, operating on f with $\hat{O}$ yields a constant, c, times f. The Schrödinger equation is, of course, an important example of an eigenvalue equation.

## 3.4 THE HARMONIC OSCILLATOR

The problem of the quantum mechanical description of a harmonic oscillator is our first example of applying the postulates of quantum mechanics. The picture of the system is the same as that in Fig. 2.1, a mass m connected by a harmonic spring with force constant k to an unmovable wall. The steps for treating this problem quantum mechanically, as well as any other problem, are the following.

i. Find the classical Hamiltonian.
ii. By replacing variables in the classical Hamiltonian with corresponding quantum mechanical operators, develop the quantum mechanical Hamiltonian operator.
iii. Find the wavefunction from the Schrödinger equation, which means finding the eigenfunction or eigenfunctions of the Hamiltonian operator.

From the development in Chap. 2, the classical Hamiltonian of the harmonic oscillator is

$$H(x,p) = \frac{p^2}{2m} + \frac{1}{2} k (x - x_o)^2$$

For convenience, let us define the x-coordinate origin to be the location of the mass when the system is at equilibrium. This means choosing the origin such that $x_o = 0$. Thus,

$$H(x,p) = \frac{p^2}{2m} + \frac{1}{2} k x^2$$

The quantum mechanical Hamiltonian is trivially constructed by formally making x, p, and H operators.

$$\hat{H} = \frac{\hat{p}^2}{2m} + \frac{1}{2} k \hat{x}^2$$

Of course, the explicit form of the momentum operator is that used earlier, and the position coordinate operator is just multiplication by x. So, a more useful way of writing the quantum mechanical Hamiltonian is

$$\hat{H} = -\frac{\hbar^2}{2m} \frac{d^2}{dx^2} + \frac{1}{2} k x^2 \quad (3\text{-}7)$$

The solution of the Schrödinger equation, $\hat{H} \psi(x) = E \psi(x)$, is the next step.

The Schrödinger equation for the harmonic oscillator happens to be a well-studied differential equation that mathematicians had solved long before the quantum mechanical problem was formulated. There are an infinite number of functions of x that turn out to be valid eigenfunctions of the Hamiltonian operator. A subscript n is used, therefore, to distinguish one solution from another. Each eigenfunction has its own energy eigenvalue, and so the differential equation at hand is

$$-\frac{\hbar^2}{2m} \frac{d^2 \psi_n(x)}{dx^2} + \frac{1}{2} k x^2 \psi_n(x) = E_n \psi_n(x) \quad (3\text{-}8)$$

All the eigenfunctions, or solutions, may be expressed as

$$\psi_n(x) = \psi_n(\tfrac{z}{\beta}) = \frac{N}{\sqrt{2^n n!}} h_n(z) e^{-z^2/2} \quad \text{where } z \equiv \beta x \quad (3\text{-}9)$$

These wavefunctions have been written as a function of z for conciseness, but z is just x scaled by the constant $\beta$. The constant $\beta$ is related to the force constant and the particle mass:

$$\beta^2 \equiv \sqrt{km} / \hbar \quad \text{and} \quad (3\text{-}10)$$

$$N = \left(\frac{\beta^2}{\pi}\right)^{1/4} \quad (3\text{-}11)$$

The n subscript has zero as its smallest allowed value and may be continued to infinity. The functions $h_n(z)$ are simple polynomials of z (or of $\beta x$) that are called Hermite polynomials.

The harmonic oscillator wavefunctions of Eq. (3-9) are all equal to a constant times a polynomial times a particular exponential function, called a Gaussian function. At z = 0 (i.e., at x = 0), the Gaussian function is at its maximum value, one. As z increases or decreases, the Gaussian function diminishes quickly. For z >> 0 or z << 0, the Gaussian function becomes vanishingly small, and so the asymptotic limit for each of the wavefunctions is zero. Restated in terms of the position variable x instead of z, the Gaussian function is

$$e^{-z^2/2} = e^{-(\beta x)^2/2}$$

Because $\beta$ increases with the mass of the particle and with the spring constant, then a heavier mass or a stiffer spring will make the value of the Gaussian function smaller for any given displacement in x.

The Hermite polynomials are very simple polynomials. It turns out that they may be generated in several ways, such as by the following formula:

$$h_n(z) = (-1)^n e^{z^2} \frac{d^n}{dz^n} e^{-z^2} \qquad (3\text{-}12)$$

Or, they may be generated by a recursive procedure.* The first several are

$$h_0(z) = 1 \qquad h_0(\beta x) = 1$$

$$h_1(z) = 2z \qquad h_1(\beta x) = 2\beta x$$

$$h_2(z) = 4z^2 - 2 \qquad h_2(\beta x) = 4\beta^2 x^2 - 2$$

$$h_3(z) = 8z^3 - 12z \qquad h_3(\beta x) = 8\beta^3 x^3 - 12\beta x$$

* The recursive relation tells how to generate the n + 1 polynomial from the $n^{th}$ order polynomial, given that $h_0(z) = 1$:

$$h_{n+1}(z) = 2\, z\, h_n(z) - \frac{dh_n(z)}{dz}$$

$$h_4(z) = 16z^4 - 48z^2 + 12 \qquad h_4(\beta x) = 16\beta^4x^4 - 48\beta^2x^2 + 12$$

$$h_5(z) = 32z^5 - 160z^3 + 120z \quad h_5(\beta x) = 32\beta^5x^5 - 160\beta^3x^3 + 120\beta x$$

Notice that $h_0$ is a constant. $h_1$ has one node, meaning that it has one point where it changes sign, namely, z = 0. Generally, $h_n$ has n nodes, and odd functions (i.e., n=1,3,5, . . . ) have one of those at the origin (x = z = 0).

A crucial point is to see that these wavefunctions are eigenfunctions of the harmonic oscillator Hamiltonian and to determine the eigenvalues associated with each. For $\psi_0$,

$$\begin{aligned}
\hat{H}\psi_0 &= -\frac{\hbar^2}{2m}\frac{d^2\psi_0}{dx^2} + \frac{1}{2}kx^2\psi_0 \\
&= -\frac{\hbar^2}{2m}\beta^2\left(\beta^2x^2 - 1\right)\psi_0 + \frac{1}{2}kx^2\psi_0 \\
&= \frac{\hbar^2}{2m}\beta^2\psi_0 + \left(\frac{k}{2} - \frac{\beta^4\hbar^2}{2m}\right)x^2\psi_0
\end{aligned}$$

A rearrangement of terms has been made between the second and third lines. This rearrangement leads not only to one term that is just a constant times the wavefunction, but also to a second term that is a different function entirely, $x^2\psi_0$. However, if the expression for $\beta$ from Eq. (3-10) is used, then the factor that multiplies $x^2\psi_0$ is found to be zero. Thus, $\psi_0$ is indeed an eigenfunction of the Hamiltonian. The eigenvalue, $E_0$, of the Schrödinger equation $\hat{H}\psi_0 = E_0\psi_0$ is

$$\begin{aligned}
E_0 &= \frac{\hbar^2\beta^2}{2m} \\
&= \frac{\hbar^2}{2m}\frac{\sqrt{km}}{\hbar} \\
&= \frac{1}{2}\hbar\sqrt{\frac{k}{m}} = \frac{1}{2}\hbar\omega
\end{aligned}$$

where $\omega$ is the intrinsic frequency of the harmonic oscillator, the same frequency as in the classical description. If the Hamiltonian is applied to the $\psi_1$ wavefunction from Eq. (3-9), the following results:

$$\begin{aligned}
\hat{H}\,\psi_1 &= -\frac{\hbar^2}{2m}\frac{d^2\psi_1}{dx^2} + \frac{1}{2}kx^2\,\psi_1 \\
&= -\frac{\hbar^2}{2m}\,\beta^2\left(\beta^2x^2 - 3\right)\psi_1 + \frac{1}{2}kx^2\,\psi_1 \\
&= 3\frac{\hbar^2}{2m}\,\beta^2\;\psi_1 + \left(\frac{k}{2} - \frac{\beta^4\hbar^2}{2m}\right)x^2\,\psi_1 \\
&= 3\frac{\hbar^2}{2m}\,\beta^2\;\psi_1 = \frac{3}{2}\hbar\omega\,\psi_1
\end{aligned}$$

The energy eigenvalue for this state is precisely $\hbar\omega$ more than the energy of $\psi_0$. And if the next state, $\psi_2$, were so tested, its eigenenergy would be $\hbar\omega$ more than the energy of $\psi_1$. In fact, the energy eigenvalues of the harmonic oscillator may be expressed generally by using the index n that distinguishes the different states.

$$E_n = (n + \frac{1}{2})\,\hbar\omega \qquad (3\text{-}13)$$

n will be referred to as a *quantum number* from now on. The state with the lowest energy is always called the *ground state*, and for the harmonic oscillator, this is the n = 0 state.

There are two very important features of Eq. (3-13) that are typical of many quantum systems. First, the allowed energies of the system are not continuous. Whereas the classical harmonic oscillator could be made to have any energy at all just by stretching it to a suitable length and releasing it, the quantum mechanical oscillator can have only the particular energies specified by Eq. (3-13). Any energy added to the oscillator or removed from it should come in chunks (quanta) of the amount that separates the allowed energies, i.e., integer multiples of $\hbar\omega$. For an oscillator in our everyday world, $\hbar$ is so small that the allowed energies are essentially continuous; this is the correspondence between the classical and quantum descriptions. The second important feature is that the

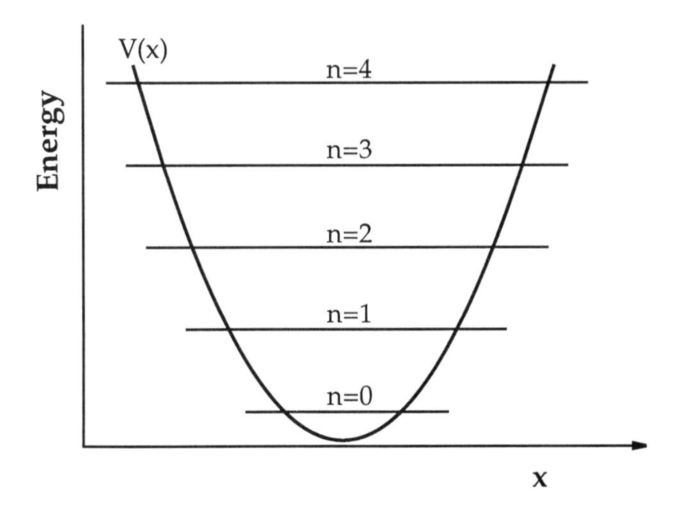


**FIGURE 3.1** The energy levels of the quantum mechanical harmonic oscillator.

ground state energy is not zero. There exists no state of the system for which the energy is zero, and so the quantum mechanical oscillator can never be completely at rest!

There is a helpful way of displaying the quantum mechanical information about the harmonic oscillator, or about other simple problems. The position coordinate, x, labels the horizontal coordinate axis of a two-coordinate graph. The vertical axis is in units of energy, and the first item to be drawn is the potential, V(x). As shown in Fig. 3.1, V(x) for the harmonic oscillator is a parabola that opens upward. Next, straight, horizontal lines are drawn at energies that correspond to the eigenenergies of the system. Each is labelled with the quantum number, n. Notice the regular spacing of these lines that is dictated by Eq. (3-13). The horizontal lines designate the *energy levels* of the system. The particular points where these lines cross the parabola, V(x), are points at which the potential energy is the same as the energy of the oscillator. Were this a classical problem, these points would be those at which the particle has no kinetic energy; it has stopped and is ready to turn back. The points are called *classical turning points,* though the quantum mechanical particle does not have to turn back at them.

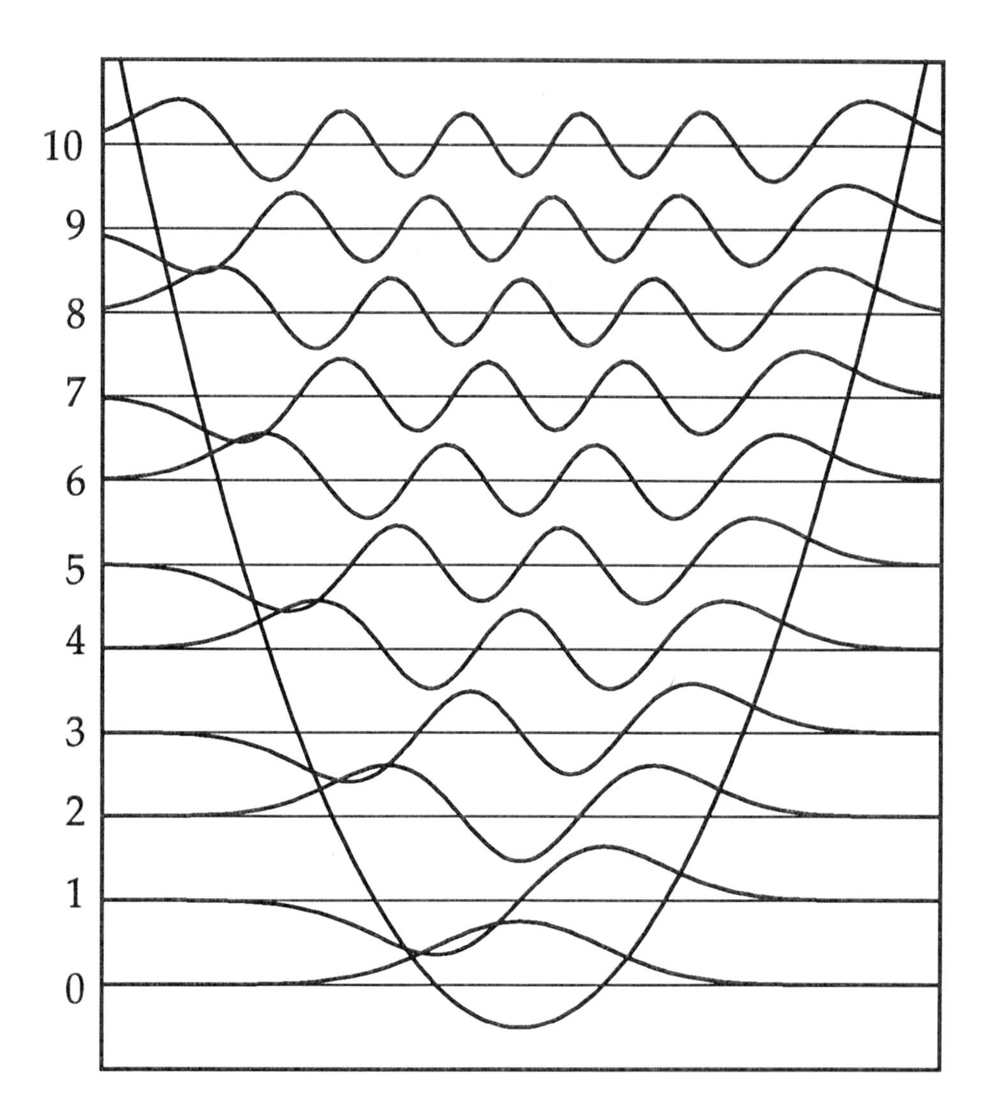


**FIGURE 3.2** The wavefunctions of low-lying levels of the quantum mechanical harmonic oscillator. The functions are drawn so that the baseline is the energy level line as in Fig. 3.1, and that means that the function is zero-valued or has a node at any point where it crosses the energy level line. Since it is the qualitative form that is usually of interest, the vertical axis is in arbitrary units. Notice that for the highest levels drawn, the wavefunction has maximum amplitude close to the classical turning points. A truly classical oscillator is moving at its slowest speed as it changes direction at a turning point, and so it is more likely to be found around the turning points than in the middle.

The horizontal lines in Fig. 3.1 that identify the energy levels serve as baselines for drawing the wavefunctions. Typically, the vertical scale for the wavefunctions is arbitrary since it is the qualitative form of each that is of most interest. Fig. 3.2 is a drawing of this sort, and one of the things we can see is that the number of nodes equals the quantum number.

## 3.5 THE PROBABILITY DENSITY

In a classical picture of electromagnetic radiation, it is the square of the wave amplitude that gives the energy density. Early notions about quantum mechanical wavefunctions of particles borrowed this idea, and it worked. Postulate IV says that *the square of the wavefunction is a function that gives the probability per unit volume, or the probability density, of finding the particles of a system to be at specific locations in space.* Designating the probability density as $\rho$, this says,

$$\rho(q_1,q_2, \dots) \equiv \psi^*(q_1,q_2, \dots)\ \psi(q_1, q_2, \dots) \qquad (3\text{-}14)$$

Notice that the "square of the wavefunction" is $\psi^*\psi$, where $\psi^*$ is the *complex conjugate* of $\psi$. The complex conjugate of a number or function is that number or function with $-i$ replacing $i$ (the square root of $-1$) wherever it shows up. While the square of a real number is the number times itself, the square of a complex number is the number times its complex conjugate. So, to take the square of a function that may be complex, we must multiply together the function and its complex conjugate.

The probability density function is as close as one can get to attaching physical significance to a wavefunction. Consider the ground state wavefunction of the harmonic oscillator. The probability density function is

$$\rho_0 = \frac{\beta}{\sqrt{\pi}}\, e^{-\beta^2 x^2}$$

This function has its maximum value at x = 0, which is the position at equilibrium. Postulate IV says that the probability per unit length (length instead of volume since this is a one-dimensional system) of finding the particle is greatest at x = 0. Loosely, the most likely spot to find the particle as it moves about is in the vicinity of x = 0.

The probability density for a number of the lowest energy states of the harmonic oscillator is shown in Fig. 3.3. These functions are sketched using the associated energy level lines as the baseline in the same way that the wavefunctions have been displayed in Fig. 3.2. At a point where the wavefunction has a node, the probability density is zero, and these points are easy to see. Notice that as the quantum number increases, the probability density becomes relatively bigger in the turning point regions than around x = 0. In this way, the quantum mechanical description is beginning to resemble the classical description. (Classically, the particle moves slowest at the turning points and spends more time in those regions.)

Should it be that the probability density function for a particle is constant in some particular region of space, then the probability of finding the particle in that region of space must result from multiplying the probability density by the volume of that region. If the probability density is not constant, then it is necessary to multiply it by an infinitesimal volume element and integrate over the region. Thus, the net probability of finding the harmonic oscillator particle between x = 0.0 and x = 0.1 is

$$\int_{0.0}^{0.1} \rho_0 \, dx$$

The probability of finding the particle *somewhere* or *anywhere* means that the limits of integration span all regions of space. For the ground state of the harmonic oscillator, this is

$$\int_{-\infty}^{\infty} \rho_0 \, dx$$

Evaluation of this integral quantity shows that it has a value of exactly one. This is understandable, since a probability of one means total

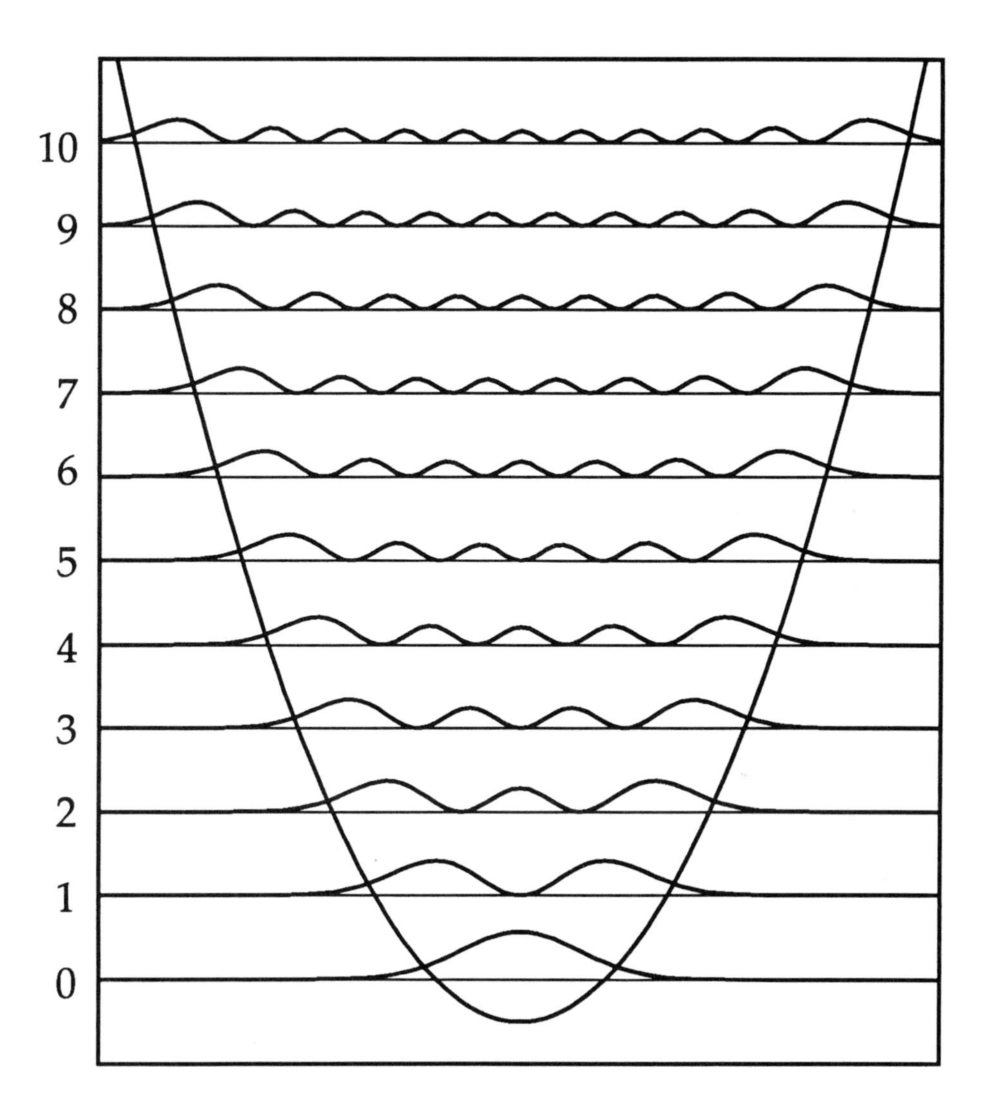


**FIGURE 3.3** The probability density functions of low-lying levels of the quantum mechanical harmonic oscillator. These functions are the squares of the wavefunctions. They are drawn so that the baseline, or point of zero probability density, is the energy level line as in Fig. 3.1. Notice that as n increases, the probability density at $x = 0$ diminishes. Furthermore, at levels such as $n = 10$ the points of maximum probability density are close to the classical turning points, unlike the lower states.

certainty, and yes, we can be certain that the particle is to be found somewhere between $x = -\infty$ and $x = \infty$.

The condition that the probability of finding the harmonic oscillator particle somewhere is one that is not implicit in the Schrödinger equation. Notice that the function $\Gamma = 3\psi_0$, for example, is just as much an eigenfunction of the Hamiltonian as $\psi_0$, and it has the same eigenvalue:

$$\hat{H}\Gamma = \hat{H}(3\psi_0) = 3\hat{H}\psi_0 = 3E_0\psi_0 = E_0(3\psi_0) = E_0\Gamma$$

But the integral of $\Gamma^*\Gamma$ over all space is 9, not 1. So, when the condition is satisfied that the integral of the probability density over all space is one, the wavefunction is in the proper form for interpretation by Postulate IV. Such a wavefunction is said to be *normalized*. A solution of the Schrödinger equation is not automatically a normalized wavefunction but may be made so by multiplying it by a constant. That constant, called a normalization factor, is obtained from the integral of the square of the unnormalized wavefunction. That is, if $\Gamma$ were an unnormalized wavefunction, then $\Gamma' = N\Gamma$ would be a normalized wavefunction given the following definition of the normalization constant, N.

$$N \equiv \frac{1}{\sqrt{\int_{-\infty}^{\infty} \Gamma^*\Gamma \, dx}} \tag{3-15}$$

If $\Gamma$ happens to be normalized already, this expression will give $N = 1$. (If the wavefunction, $\Gamma$, were a function of several coordinates, $q_1$, $q_2$, $q_3$, and so on, then $dq_1\, dq_2\, dq_3 \ldots$ would be in place of dx in the integral.)

Postulate II says that a measurement of a dynamical variable equals an eigenvalue *if* the wavefunction is an eigenfunction of the operator associated with that variable. In the case of the harmonic oscillator, it is reasonable to ask what would result from measuring the position of the particle, x, even though none of the wavefunctions are eigenfunctions of the operator x. Postulate V says that in such a case only the average of many measurements can be determined from the wavefunction. Individual measurements may differ from one to the next. The average is obtained by integrating the variable's probability density, which is just $\psi^*\psi$ with the associated operator sandwiched in between. According to this postulate *the average or mean value of measuring the quantity associated with some operator $\hat{T}$ for a system described by a normalized wavefunction $\Psi$ is*

$$\int \Psi^* \hat{T} \Psi \, d\tau$$

$d\tau$ is the volume element for the coordinate system in which $\Psi$ is given. This average value is also called the *expectation value* of T and is usually designated as <T>. If subscripts are attached to this symbol, they are used to indicate the quantum number(s) of the particular wavefunction used in the integral.

For the harmonic oscillator, it is directly established* that for all n,

$$\langle x \rangle_n = \int_{-\infty}^{\infty} \psi_n^* \, x \, \psi_n \, dx = 0$$

For all states of the harmonic oscillator, the expectation value of the position of the particle is zero, which is the location of the potential minimum or the middle of the potential well. This is a consequence of the fact that for every state, the probability density at some x is exactly the same as at $-x$. The probability density function is symmetric with respect to the middle of the well, and there is the same chance of being a certain amount to the left of the middle as being to the right by that amount. An average of many measurements gives zero, although any one measurement need not give zero. In fact, a series of measurements may give many different values that are distributed in some range.

The *standard deviation* is a statistical estimator of the range of most measurements in some series of trials. It is the root mean square (rms) deviation from the average value. Let's say a particular set of measurements of some quantity yielded these five values, 1.2, 2.4, 2.0, 1.8, and 1.6, the average of these being 1.8. The deviations of each measurement from this average are –0.6, 0.6, 0.2, 0.0, and –0.2, respectively. The deviations squared are 0.36, 0.36, 0.04, 0.0, and 0.04, and the mean of these values is 0.16. The rms deviation is $\sqrt{0.16} = 0.4$. Three of the five

* The function in the integrand, $f(x) = \psi_n(x)^* \, x \, \psi_n(x)$, must be an odd function because it is the product of an odd function, x, and an even function, $\psi_n(x)^* \psi_n(x)$. This means that for any value of the variable x, $f(x) = -f(-x)$. When an odd function is integrated from $-\infty$ to $+\infty$, the result must be zero.

measurements happen to be in the range 1.8 ± 0.4.

The same type of deviation in measurement from an average value may be evaluated from the wavefunction in a quantum mechanical description of a system. First, there is an operator that corresponds to "measuring the deviation from the average" of some dynamical variable. If that variable is T this operator is simply $\hat{T} - \langle T \rangle$. It is the T operator less the numerical value that is the expectation value of T. The square of this operator, $(\hat{T} - \langle T \rangle)^2$, corresponds to measuring the square of the deviation from the average. Thus, the mean square of the deviation is the expectation value of this operator,

$$(\Delta T)^2 \equiv \int \Psi^* (\hat{T} - \langle T \rangle)^2 \, \Psi \, d\tau \qquad (3\text{-}16)$$

and the root mean square deviation is the square root of this number, $\Delta T$. It is called the *uncertainty* in the measurement of the variable.

From Eq. (3-16), another expression may be developed for $\Delta T$.

$$\int \Psi^* (\hat{T} - \langle T \rangle)^2 \, \Psi \, d\tau = \int \Psi^* (\hat{T}^2 - 2\hat{T} \langle T \rangle + \langle T \rangle^2) \, \Psi \, d\tau$$

$$= \int \Psi^* \hat{T}^2 \, \Psi \, d\tau - 2 \langle T \rangle \int \Psi^* \hat{T} \, \Psi \, d\tau + \langle T \rangle^2 \int \Psi^* \Psi \, d\tau$$

$$= \langle T^2 \rangle - 2 \langle T \rangle^2 + \langle T \rangle^2 = \langle T^2 \rangle - \langle T \rangle^2$$

Therefore,

$$\Delta T = \sqrt{\langle T^2 \rangle - \langle T \rangle^2} \qquad (3\text{-}17)$$

The uncertainty in T is the square root of the difference between the expectation value of the operator "T squared" and the square of the expectation value of T.

*Heisenberg's uncertainty relation* states that the product of the measurement uncertainties of two conjugate variables (see Chap. 2) is a number on the order of Planck's constant or larger. (Of course, $\hbar$ is a very tiny value in our macroscopic everyday world.) Position and momentum are conjugate variables, and so, if one may be measured with relatively little uncertainty, then the other cannot.

If a wavefunction is an eigenfunction of an operator of interest, the uncertainty is identically zero, which is consistent with Postulate II. For instance, if the operator $\hat{T}$ applied to some function $\Gamma$ yields the eigenvalue t times $\Gamma$, then the following result is obtained.

$$\hat{T}^2 \Gamma = \hat{T}(\hat{T}\Gamma) = \hat{T}\, t\, \Gamma = t^2 \Gamma \quad \Rightarrow \quad \langle T^2 \rangle = t^2 = \langle T \rangle^2$$

This means $\Delta T$ is zero. There is no uncertainty in the measurement of T in this case. Every measurement will yield the same value, namely, t. This also means that the uncertainty in the variable conjugate to T must then be infinite according to Heisenberg's uncertainty relation. In later sections, we shall see the circumstances for energy and time to be taken as conjugate variables, and then if a system's energy has no uncertainty, its lifetime (the uncertainty in time) will be infinite.

## 3.6 THE PARTICLE-IN-A-BOX PROBLEM

The particle-in-a-box is a special model problem in quantum mechanics. It is the problem of a hypothetical particle of mass m that may move in one dimension, along the x-axis, and that experiences a potential, V(x), with a very simple form. V(x) is zero in some region, $0 < x < l$, and is infinite elsewhere. This is not a realistic situation by any means because realistic potentials do not become infinite throughout a region and do not change abruptly from zero to infinite. However, certain potential functions may be relatively flat over a region and turn sharply upward at the ends of the region. In such a case, V(x) would serve as an approximation, as shown in Fig. 3.4.

We assume that even quantum particles will not be found in regions where the potential is infinite. Then, V(x) is really a "box"; the particle will be found only in the region $0 < x < l$. Within that region, the potential is zero, and so the time-independent Schrödinger equation is extremely simple,

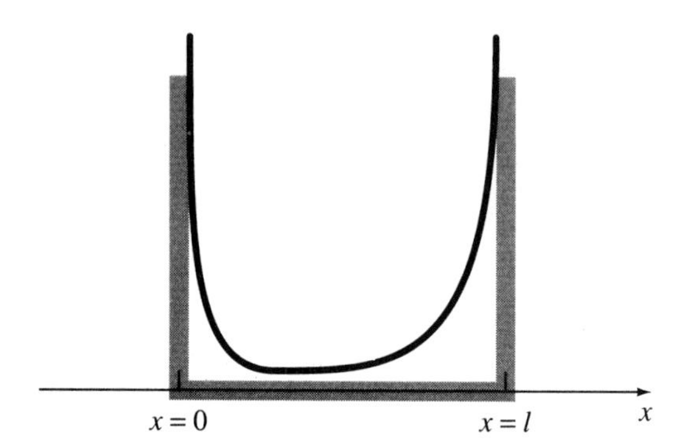


**FIGURE 3.4** The particle-in-a-box potential, which is an impenetrable potential outside x = 0 and x = $l$ and zero inside, might serve as an approximation of the smoothly changing potential represented by the solid line.

$$-\frac{\hbar^2}{2m}\frac{d^2\psi(x)}{dx^2} = E\,\psi(x) \qquad (3\text{-}18)$$

The general solution of this differential equation may be written two ways.

$$\psi(x) = A\,e^{i\sqrt{2mE}\;x/\hbar} + B\,e^{-i\sqrt{2mE}\;x/\hbar} \qquad (3\text{-}19a)$$

$$\psi(x) = a\sin\left(\sqrt{2mE}\;x/\hbar\right) + b\cos\left(\sqrt{2mE}\;x/\hbar\right) \qquad (3\text{-}19b)$$

The constants, A and B, or a and b, and the energy eigenvalues do not end up being specified in the course of finding the general solution. Of course, the Schrödinger equation that was used is only valid in a certain region, the confines of the box, and so certain information about the problem has been taken into account.

The particle-in-a-box potential is discontinuous at x = 0 and at x = $l$. This presents a complication in the way we may approach the problem because we cannot write one Hamiltonian that is correct for the region $x < 0$, and for the region $0 < x < l$, and for the region $x > l$. In general, discontinuous potentials require breaking up the problem into regions.

(The boundaries of the regions are wherever the potentials are discontinuous.) Separate Schrödinger equations may be solved for each region, but there are overriding conditions that must be met in order for us to interpret the squares of wavefunctions as probability densities. The conditions are that the wavefunctions are single-valued everywhere, that they are continuous, and that their first derivatives are continuous. Within each region, these conditions are automatically satisfied. At the boundaries, these conditions have to be imposed.

In the particle-in-a-box problem, the wavefunction must be zero-valued at the edges and outside the confines of the box. Thus, the allowed solutions within the box, Eq. (3-19), are only acceptable as wavefunctions if they become zero-valued at the edges of the box. This is expressed as

$$\psi(0) = 0 \quad \text{and} \quad \psi(l) = 0$$

Using the trigonometric form of the solutions in the box, the first of these conditions forces the constant b to be zero. The second condition leads to the following result.

$$\psi(l) \;=\; a \sin\left(l\sqrt{2mE}\;/\;\hbar\right) \;=\; 0 \qquad \Rightarrow$$

$$l\sqrt{2mE}\;/\;\hbar \;=\; n\,\pi \qquad \text{where } |n| = 0, 1, 2, 3, \ldots \tag{3-20}$$

Rearranging this result shows it to be a condition on the energy:

$$E \;=\; \frac{n^2\,\pi^2\,\hbar^2}{2m\,l^2} \tag{3-21}$$

The allowed energies, then, are not continuous because of the boundary potential. In fact, particles that are bound or trapped by potentials have discrete energy levels. Notice that the separation of adjacent energy levels increases with n because of the quadratic dependence of the energy eigenvalues on n in Eq. (3-21). In the harmonic oscillator, the bound states were separated by an equal amount; the eigenenergies varied linearly with the quantum number.

Using the relation in Eq. (3-20) allows us to express the wavefunctions as

$$\psi_n(x) \quad = \quad a \sin\left(\frac{n\pi x}{l}\right) \tag{3-22a}$$

At this point, we reject the possibility of n = 0 as unphysical. It would correspond to a wavefunction that is zero-valued everywhere and has a zero probability density everywhere. Also, we exclude the negative integers as possible quantum numbers. Even though they are allowed by Eq. (3-20), they do not generate unique wavefunctions. Instead, they correspond to a wavefunction with a sign or phase change since

$$\sin\left(\frac{-n\pi x}{l}\right) \quad = \quad -\sin\left(\frac{n\pi x}{l}\right)$$

Thus, the allowed values of n are 1, 2, 3, and so on. Finally, the normalization condition determines the constant a, which turns out be independent of n:

$$\int_0^l \psi_n^* \psi_n \, dx \;=\; \int_0^l a^2 \sin^2(n\pi x/l)\, dx \;=\; a^2 l/2 \;\Rightarrow\; 1$$

The wavefunctions for the particle-in-a-box problem are

$$\psi_n(x) \quad = \quad \sqrt{\frac{2}{l}} \sin\left(\frac{n\pi x}{l}\right) \tag{3-22b}$$

with the wavefunction being zero outside the box.

## 3.7 PARTICLES AND POTENTIALS

The Schrödinger equation for a particle free to move in one dimension with no potential is

$$-\frac{\hbar^2}{2m}\frac{d^2\psi(x)}{dx^2} \quad = \quad E\,\psi(x) \tag{3-23}$$

The solutions of this differential equation may be obtained by simple integration.

$$\psi(x) = A e^{-i\sqrt{2mE}\, x/\hbar} + B e^{i\sqrt{2mE}\, x/\hbar} \quad (3\text{-}24)$$

where A and B are constants. Notice that there is no restriction on the energy E. Any value is allowed because the particle is free. Sometimes the constant in the exponential is collected into one constant, k, called the *wave constant* for the particle.

$$k = \sqrt{2mE} / \hbar \quad (3\text{-}25)$$

$$\psi(x) = A e^{-ikx} + B e^{ikx} \quad (3\text{-}26)$$

If the particle is free to move in three dimensions, the Schrödinger is, of course, separable, and the wavefunctions are products of independent functions of x, y, and z that have the form given in Eq. (3-24) or (3-26). However, the wave constants are not necessarily the same in each direction; there will be a $k_x$, a $k_y$, and a $k_z$ in the wavefunction, and these are independent. Collected into a vector, they comprise the *wave vector* for the moving free particle.

$$\vec{k} = (k_x, k_y, k_z) \quad (3\text{-}27)$$

The energy is partitioned in the three directions because of the separability of the Schrödinger equation.

$$-\frac{\hbar^2}{2m}\nabla^2 \Psi(x,y,z) = \frac{\hbar^2}{2m}(k_x^2 + k_y^2 + k_z^2)\, \Psi(x,y,z)$$

$$= (E_x + E_y + E_z)\, \Psi(x,y,z) \quad (3\text{-}28)$$

This is equivalent to expressing the kinetic energy of a free classical particle with velocity components as $m(v_x^2 + v_y^2 + v_z^2)/2$.

The wavefunction of the free particle in three dimensions is an eigenfunction of $p_x^2$, $p_y^2$ and $p_z^2$. It may also be an eigenfunction of $p_x$ or $p_y$ or $p_z$. The general solution of the differential equation, or the function that satisfies Eq. (3-28), is the product of functions of the form of Eq. (3-26).

$$\Psi(x,y,z) = \left( A_x e^{-ik_x x} + B_x e^{ik_x x} \right)\left( A_y e^{-ik_y y} + B_y e^{ik_y y} \right)$$

$$\times \left( A_z e^{-i k_z z} + B_z e^{i k_z z} \right) \tag{3-29}$$

If either $A_x$ were zero or $B_x$ were zero, the wavefunction would be an eigenfunction of $p_x$. If either $A_y$ were zero or $B_y$ were zero, the wavefunction would be an eigenfunction of $p_y$, and likewise for the z-direction. This is seen from the following.

$$\hat{p}_x \left( A_x e^{-i k_x x} \right) = -i\hbar A_x \frac{\partial}{\partial x} e^{-i k_x x} = -\hbar k_x \left( A_x e^{-i k_x x} \right)$$

The eigenvalue for the momentum is $-\hbar k_x$. Furthermore, there is no uncertainty in the momentum component because this function is also an eigenfunction of $p_x^2$. If $A_x$ were zero instead, the eigenvalue would be $\hbar k_x$. The positive eigenvalue implies that the momentum is that associated with motion in the +x direction, whereas the negative eigenvalue is associated with motion in the opposite direction. Thus, the complex exponential forms of the wavefunctions of a free particle may be associated with motion to the right or to the left along each coordinate axis. Thus, the set of elements termed the wave vector, $\vec{k}$, in Eq. (3-27) seems appropriately named. We may now see it as a vector that identifies the direction that a free particle is moving in three-dimensional space.

The wavefunctions of the free particle are said to form a *continuum of energy states* because a continuous choice of energies (or wave vectors) is possible. Unlike the harmonic oscillator or the particle in the box, there is no restriction on what energies are allowed. This is a typical feature of a particle that is unbound, where "unbound" means that the particle's kinetic energy is sufficient to surmount any potential energy barrier in the direction of its motion. A classical particle meeting this description would continue in that direction endlessly.

An interesting feature of the harmonic oscillator wavefunctions was that the probability density is not zero beyond the classical turning points, though it is small. Since the classical turning points are the points at which the particle's total energy equals the potential energy, then the nonzero possibility of finding the particle in a region where the potential is still greater is remarkable. This is the quantum phenomenon called *tunneling,*

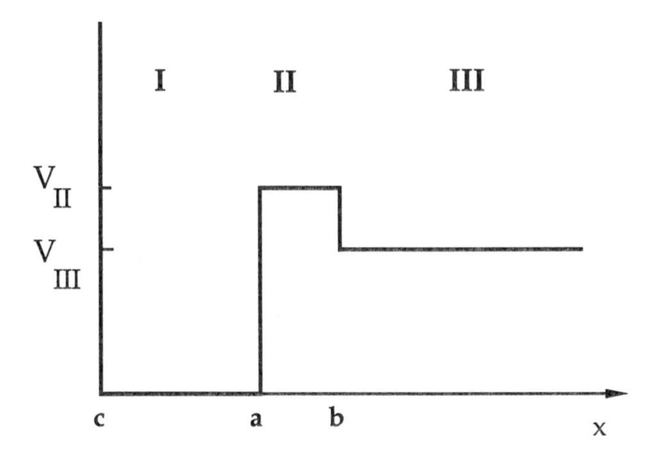


**FIGURE 3.5** A hypothetical step potential for a one-dimensional particle. To understand this system, it is helpful to break it into regions. In the first region (I), there is a potential well that is almost like the potential of the particle-in-a-box problem. The second region (II) has a constant, but not infinite potential. The third region (III) has a flat potential, and this continues to infinity.

which is a nonzero probability density for finding a particle in a region where the potential is greater than the particle's energy.

Tunneling becomes particularly interesting in the situation where the potential has the form shown in Fig. 3.5. If a particle's energy, E, were in the range $V_{III} < E < V_{II}$, and if it were somehow placed into region I, we might expect some tunneling from region I into region II. And then, as long as the wavefunction did not diminish completely to zero as it continued to the right in region II, there would be some possibility that it would enter region III. There, it would behave as a free particle with kinetic energy equal to $E - V_{III}$. This tunneling occurrence would make a trapped particle free, and that is something that in the classical world would be the same thing as walking through a brick wall. Tunneling through a barrier is an immensely important physical process. It explains spontaneous nuclear fission and beta decay, and in chemistry, it explains why certain reactions take place even when the reactants have insufficient energy to surmount the activation barrier.

The probability that tunneling will take place can be deduced from the wavefunctions. The step potential in Fig. 3.5 serves as a good example of the possibility of tunneling and also of the quantum mechanical complications of potentials that change sharply at some point. So, the next task is to understand something about the wavefunctions for the problem represented by Fig. 3.5.

The connection or continuity conditions at region boundaries are of central importance in finding the wavefunctions for step potentials. First, we may assume a free-particle type of wavefunction within each region because the potential is unchanging, or is flat. For the problem represented by Fig. 3.5, this means

$$\psi_I(x) = A_I e^{-ik_I x} + B_I e^{ik_I x} \quad \text{where } k_I = \sqrt{2mE}\,/\hbar \tag{3-30}$$

$$\psi_{II}(x) = A_{II} e^{-ik_{II} x} + B_{II} e^{ik_{II} x} \quad \text{where } k_{II} = \sqrt{2m(E - V_{II})}\,/\hbar \tag{3-31}$$

$$\psi_{III}(x) = A_{III} e^{-ik_{III} x} + B_{III} e^{ik_{III} x} \quad \text{where } k_{III} = \sqrt{2m(E - V_{III})}\,/\hbar \tag{3-32}$$

Then, for the wavefunction to be continuous at the boundary regions, the following conditions must be satisfied.

$$\psi_I(a) = \psi_{II}(a) \tag{3-33}$$

$$\psi_{II}(b) = \psi_{III}(b) \tag{3-34}$$

For the wavefunction to have a continuous first derivative everywhere, the following conditions must be satisfied as well.

$$\left.\frac{d\psi_I}{dx}\right|_{x=a} = \left.\frac{d\psi_{II}}{dx}\right|_{x=a} \tag{3-35}$$

$$\left.\frac{d\psi_{II}}{dx}\right|_{x=b} = \left.\frac{d\psi_{III}}{dx}\right|_{x=b} \tag{3-36}$$

Also, the wavefunction must vanish at x = c because the potential is infinite at that point:

$$\psi_I(c) = 0 \tag{3-37}$$

Therefore, we have five connection conditions on the six constants $A_I$, $A_{II}$, $A_{III}$, $B_I$, $B_{II}$, and $B_{III}$. Though these are insufficient to completely determine the six constants, they do imply relationships among them.

Let us take as an initial condition on the system in Fig. 3.5 that each and every particle came streaming in from the far right (i.e., x > b) with an energy, $E > V_{III}$. The quantum mechanical statement of this condition is that in region III, the eigenvalue associated with momentum must be negative, since a negative eigenvalue corresponds to motion in the –x direction which is to the left. The momentum operator is $-i\hbar\, d/dx$, and so $\psi_{III}$ will have a negative eigenvalue with the momentum operator only if $B_{III}$ equals zero. Thus, with the initial condition that a particle encountered the potential in Fig. 3.5 by approaching from the far right, we may say that

$$\psi_{III} = A_{III}\, e^{-i k_{III} x}$$

The first boundary to consider is at x = b, the boundary between regions II and III. Since a coordinate origin has not been specified, we may choose it to be at this boundary; that is, b = 0. The continuity conditions of Eqs. (3-34) and (3-36) are now

$$A_{III} = A_{II} + B_{II}$$

$$-k_{III} A_{III} = -k_{II} A_{II} + k_{II} B_{II}$$

By rearranging, we can obtain $A_{II}$ and $B_{II}$ in terms of $A_{III}$.

$$A_{II} = \frac{A_{III}}{2}\left(1 + \frac{k_{III}}{k_{II}}\right) \tag{3-38}$$

$$B_{II} = \frac{A_{III}}{2}\left(1 - \frac{k_{III}}{k_{II}}\right) \tag{3-39}$$

The ratio, $k_{III}/k_{II}$, is the key factor in the relative size of the amplitudes, $A_{II}$ and $B_{II}$.

The initial condition that we are using is that of particles approaching from the far right. This is implied by $B_{II}$ being zero. In the same way, we may say that if $B_{II}$ were zero, then the wavefunction in region II would correspond to motion to the left; or, if $A_{II}$ were zero, the wavefunction would correspond to motion to the right. When neither coefficient is zero, the postulates provide an interpretation of the probability of motion being to the left or to the right. At any point in region II, the normalized probability that a measurement of the momentum will yield a negative value (motion to the left) is

$$P_{II}^{left} = \frac{A_{II}^{*} A_{II}}{A_{II}^{*} A_{II} + B_{II}^{*} B_{II}} \tag{3-40}$$

The probability of motion in the opposite direction is

$$P_{II}^{right} = \frac{B_{II}^{*} B_{II}}{A_{II}^{*} A_{II} + B_{II}^{*} B_{II}} \tag{3-41}$$

And clearly, the probability of motion in either direction is properly one.

Given the initial condition of the particle approaching from the far right of region III, we may regard the probability in Eq. (3-40) as a measure of the likelihood of *transmission*, T, since it gives the likelihood that the particle will continue in the same direction it was going initially. We may regard the probability in Eq. (3-41) as a measure of the likelihood of *reflection*, R, since it is the probability that the particle has reversed direction or has been reflected back. From Eqs. (3-38) and (3-39), we may substitute to obtain the following expressions.

$$T \equiv P_{II}^{left} = \frac{(1+g)^2}{(1+g)^2 + (1-g)^2} = \frac{(1+g)^2}{2\,(1+g^2)}$$

$$R \equiv P_{II}^{right} = \frac{(1-g)^2}{2\,(1+g^2)}$$

where $g = k_{III}/k_{II}$, and where it has been assumed that $E > V_{II}$. Next, using Eqs. (3-31) and (3-32), g can be expressed in terms of E, the particle's energy, $V_{II}$, and $V_{III}$. This leads to the following result.

$$T = \frac{1}{2} + \frac{\sqrt{(E - V_{II})(E - V_{III})}}{2E - V_{II} - V_{III}} \tag{3-42}$$

$$R = \frac{1}{2} - \frac{\sqrt{(E - V_{II})(E - V_{III})}}{2E - V_{II} - V_{III}} \tag{3-43}$$

From this, we see that the likelihood of reflection and transmission at a step potential boundary can have a simple dependence on the difference between the particle's energy and the heights of the two potential steps.

To understand the implications of Eqs. (3-32) and (3-43), we shall consider certain specific choices of the potentials and the particle's energy.

1. If $V_{II}$ were equal to $V_{III}$, then the rightmost terms in Eqs. (3-42) and (3-43) would reduce to 1/2. Then, we would have T = 1 and R = 0. This is reassuring for the condition that $V_{II}=V_{III}$ means that there would be no potential step (i.e., no boundary) and so there would be complete transmission.
2. If the particle's energy, E, were much, much greater than both $V_{II}$ and $V_{III}$, then the rightmost terms would be very nearly equal to 1/2. Again, that would lead to T = 1 and R = 0. Classically, the difference, $E - V_{II}$ (or $E - V_{III}$), is the particle's kinetic energy in the region. So, if it has a lot of kinetic energy, the step in the potential is not too important; mostly the particle is likely to continue ahead. This is illustrated in Fig. 3.6.
3. If E were much, much greater than $V_{III}$, but only 10% bigger than $V_{II}$, then the rightmost terms in Eqs. (3-42) and (3-43) could be approximated:

$$\frac{\sqrt{(E - V_{II})(E - V_{III})}}{2E - V_{II} - V_{III}} \cong \frac{\sqrt{(E - V_{II})E}}{2E - V_{II}} = \frac{\sqrt{(1.1V_{II} - V_{II})\,1.1\,V_{II}}}{2\,(1.1\,V_{II}) - V_{II}}$$

$$= \sqrt{0.11}\;/\;1.2 \;=\; 0.28$$

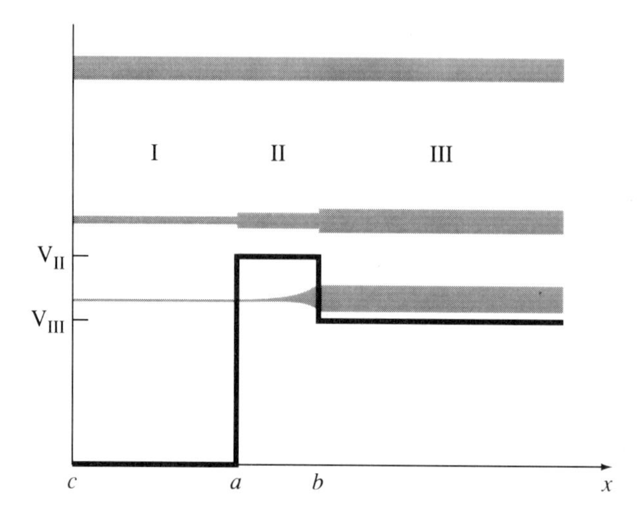


**FIGURE 3.6** The transmission of a beam of particles coming from the right and encountering the potential of Fig. 3.5 is represented by three horizontal lines corresponding to three specific particle energies. The width of each line is a representation of the transmission probability as a function of x. In the highest energy case, there is a negligible diminishment at the boundary between regions II and III, but at lower energies, there is a noticeable diminishment.

Then, T = 0.78 and R = 0.22, and in contrast to the prior specific case, there is a significant probability that the incoming particle will be reflected back. This is at odds with the experience in our macroscopic, classical mechanical world where an object encountering a potential will not be stopped or reflected if it has energy in excess of the potential energy. In the quantum world, there is a probability of being reflected even if a particle's energy is more than the potential barrier. That probability grows as the particle's energy diminishes, and were E equal to $V_{II}$, then T and R would both be 1/2. This is illustrated in Fig. 3.6.

The important conclusion from this analysis is that a quantum mechanical particle may be reflected by encountering a potential barrier even if it has sufficient energy to pass over the barrier.

Let us next consider the case where the particle is approaching from the far right, just as before, but that the energy happens to be such that $V_{III} < E < V_{II}$. For this case, $k_{II}$ will be imaginary [see Eq. (3-31)], and so the general solution to the Schrödinger equation for region II is best expressed as

$$\psi_{II}(x) = A_{II} e^{-\alpha_{II} x} + B_{II} e^{\alpha_{II} x} \tag{3-44}$$

$$\text{where } \alpha_{II} = \sqrt{2m(V_{II} - E)} / \hbar$$

The connection conditions lead to relations similar to Eqs. (3-38) and (3-39).

$$A_{II} = \frac{A_{III}}{2}\left(1 + i\frac{k_{III}}{\alpha_{II}}\right) \tag{3-45}$$

$$B_{II} = \frac{A_{III}}{2}\left(1 - i\frac{k_{III}}{\alpha_{II}}\right) \tag{3-46}$$

Notice that $B_{II}$ is the complex conjugate of $A_{II}$. If we substitute this result into Eq. (3-44) and then group the real and imaginary terms, we obtain the following expression for the wavefunction.

$$\psi_{II}(x) = \frac{A_{III}}{2}\left(e^{-\alpha_{II} x} + e^{\alpha_{II} x}\right) + i\frac{A_{III} k_{III}}{2\alpha_{II}}\left(e^{-\alpha_{II} x} - e^{\alpha_{II} x}\right)$$

$$= A_{III}\left[\cosh(\alpha_{II} x) - i\frac{k_{III}}{\alpha_{II}} \sinh(\alpha_{II} x)\right] \tag{3-47}$$

This wavefunction is complicated, but the important result is that it is not strictly zero. That means the particle penetrates into region II even though it does not have the energy to surmount the potential barrier ($E<V_{II}$). Were it a classical particle, there would be no chance of finding it in region II, and so this quantum feature is referred to as tunneling through a barrier. The particle is said to be in a *classically forbidden* region, somewhat like passing through a solid wall. We do not develop reflection and transmission probabilities in this situation since this is not a free particle and the wavefunction does not consist of simple eigenfunctions of the momentum operator.

Model problems that we have considered illustrate several features of quantum systems that are outside our experience in our macroscopic world. (i) A particle trapped in a potential, meaning the potential rises higher than the particle's energy, has discontinuous allowed energies. The basic examples are the harmonic oscillator and the particle in a box. If the particle is not trapped, then there is a continuum of allowed energy states. (ii) A particle may tunnel into a region where it is classically forbidden, meaning where its energy is below the potential energy. The likelihood for tunneling diminishes as the difference in the potential and the particle's energy grows. (iii) The likelihood that a particle will pass through a barrier region (e.g., a step potential) diminishes with increasing width of the region. (iv) There is a probability that a particle will be reflected by a barrier even if its energy is sufficient to surmount the barrier.

## Exercises

1. Ignoring relativistic effects calculate the de Broglie wavelength for a free electron with a velocity 10% of the speed of light. How slow would a 10.0 g marble be moving if it had the same de Broglie wavelength?
2. From the definition of operator multiplication and the definition of a commutator, find a single operator that is the same as the following operators.

   *a.* $[\hat{B}, \hat{A}^2]$ *b.* $[\hat{B}^2, \hat{A}]$ *c.* $[\hat{B}^2, \hat{A}^2]$ *d.* $[\hat{A}^2, \hat{B}^2]$

   where $\hat{A} = x$ and $\hat{B} = d/dx$.
3. Evaluate $[e^{-3x}, d^2/dx^2]$ and $[e^{-x}, d^2/dx^2]$.
4. Consider a 10.0 g mass attached to a spring with a force constant such that the oscillation frequency is 1 cycle/sec. This is a system that we could encounter in our macroscopic world, but treating it quantum mechanically, what is the energy separation between different allowed energy states and what is the zero-point energy? Is the quantization of energy consistent with our perceptions of continuous allowed energies for classical (macroscopic) oscillators? What would

be the n quantum number of this oscillator if it had an energy of about 0.001 ergs?

5. Starting from Eq. (3-9), write out the first five harmonic oscillator wavefunctions explicitly in terms of x.

6. Apply the harmonic oscillator Hamiltonian to $\psi_2$ and to $\psi_3$ and verify that they are eigenfunctions.

7. Establish that $\psi_0$ of the harmonic oscillator is an eigenfunction of the following operator, and deduce why it must be.

$$\hat{A} = \frac{\hbar^4}{4m^2}\frac{d^4}{dx^4} - \frac{\hbar^2\omega^2}{2}\left(x^2\frac{d^2}{dx^2} + 2x\frac{d}{dx} + 1\right) + \frac{k^2x^4}{4}$$

8. Find the expectation value of the square of the momentum for the n = 0 and n = 1 states of the harmonic oscillator.

9. At what value or values of x is the probability density function of the n = 1 wavefunction of the harmonic oscillator at maximum value?

10. Find $\Delta p$ for the n = 0 and n = 1 states of the harmonic oscillator.

11. Find the expectation value of $x^2$ for the n = 0, n = 1, n = 2 and n = 3 harmonic oscillator wavefunctions. How does this quantity depend on the quantum number n?

12. Given that the expectation value $< x >$ is zero for all harmonic oscillator states, use the result of Prob. 11 to develop an expression that gives $\Delta x$ in terms of the n quantum number (i.e., that gives $\Delta x$ as a function of n).

13. For a hypothetical harmonic oscillator with a mass equal to the mass of a hydrogen atom and a force constant such that the frequency of vibration is $1.0 \times 10^{14}$ cycles/sec, find the classical turning points for the n = 0, n = 1, n = 2, and n = 3 levels. Then calculate the probability density for each wavefunction at one of its turning points. On the basis of these values, what would you expect for levels with very large n quantum numbers?

14. Use the recursion relation for Hermite polynomials to find $h_7(z)$.

15. Evaluate $\psi_n(0)$ (i.e., find the numerical value of the $n^{th}$ wavefunction at x = 0) for the n=0, 2, 4, and 6 (normalized) states of the harmonic oscillator. Is there an apparent trend in these values? Compare with the value of $\psi_n(0)$ where n is an odd integer.

16. Find the expectation value <x> and the uncertainty $\Delta x$ for the lowest three states of the one-dimensional particle in a box.

17. Express the one-dimensional particle-in-a-box wavefunctions of Eq. (3-22b) as a superposition of complex exponential functions. The general form was given in Eq. (3-19b). With this equivalent form, find the expectation value <p> and the uncertainty $\Delta p$ for the lowest three states of the one-dimensional particle in a box.

18. Make a sketch of the energy levels of the one-dimensional particle in a box by drawing a straight horizontal line for each level, placing the lines against a vertical energy scale. Assume the length of the box is $l$. To the left, sketch the same energy levels for the particle in a box if the length is $l/2$, and to the right sketch the same energy levels if the length of the box is $2l$.

19. Based on the wavefunctions for the harmonic oscillator and the wavefunctions for the particle in a box, construct a sketch to show the qualitative features of the three lowest energy wavefunctions for a particle trapped in a potential, V(x), that is infinite at x = 0, and that for $x > 0$ is $kx^2/2$ where k is a constant.

20. If the particle of the one-dimensional particle-in-a-box problem were an electron, what would the length of the box, $l$, have to be in order for there to be a transition energy of 10,000 $cm^{-1}$ for a transition from the lowest state to the first excited state? What would the value of n for $H-(C{\equiv}C)_n-H$ have to be for a linear polyene to have approximately this length?

## Bibliography


1. W. Kauzman, *Quantum Chemistry* (Academic Press, New York, 1957). The correspondence between classical vibration and the wave nature of quantum mechanical systems was very important in the early understanding of quantum phenomena. The first chapters of this text highlight that.

2. J. C. Davis, *Advanced Physical Chemistry* (Ronald Press, New York, 1965).
3. M. Karplus and R. N. Porter, *Atoms and Molecules* (Benjamin, Menlo Park, California, 1970).
4. F. L. Pilar, *Elementary Quantum Chemistry* (McGraw-Hill, New York, 1990).
5. D. A. McQuarrie, *Quantum Chemistry* (University Science Books, Mill Valley, California, 1983).

# *Chapter 4*

# Quantum Mechanics - II

*With the basic postulates of quantum mechanics in hand, we now consider some of the formal elements of quantum mechanical problems and the mathematics for solving them. Variation theory and perturbation theory, introduced in this chapter, are two very important tools for solving or for approximately solving the Schrödinger equations that arise for many different types of problems. Angular momentum, a special aspect of quantum mechanical systems, can often be treated separately from solution of the Schrödinger equation, and this is discussed.*

## 4.1 HERMITIAN OPERATORS

The manipulation of wavefunctions and operators can be presented with more concise notation than has been used so far. One way, due to Dirac, is called "bra-ket" notation. First, the integration symbols, $\int \ldots d\tau$, are replaced with angle brackets, $< \ldots >$. Thus, anything placed within these brackets is meant to be integrated over all spatial coordinates of the system. Between the brackets there is a vertical bar that separates functions, as in

$<f(x) \mid g(x)>$. It is implicit that the complex conjugate is to be used for anything between the left bracket and this vertical bar, and so $<f(x) \mid g(x)> = \int f^*(x)\, g(x)\, d\tau$. Should we happen to write a function to the left of the bar that is explicitly the complex conjugate of some function, it remains *implicit* to take its complex conjugate: $<f^*(x) \mid g(x)> = \int f(x) g(x)\, d\tau$.

An operator acting on a function on the left side of the vertical bar does not act on anything to the right of the vertical bar. An operator to the right of the vertical bar acts only on functions to its right, as would be natural from operator algebra. Thus, an expectation value for the wavefunction $\Psi$ and the operator $\hat{O}$ is written† $<\Psi \mid \hat{O}\Psi>$, but this means exactly the same as $\int \Psi^* \hat{O} \Psi\, d\tau$. Likewise, $<\hat{O}\Psi \mid \Phi> \equiv \int \Phi\, (\hat{O}^* \Psi^*)\, d\tau$ because the operator acts only on the function $\Phi$ by this notation.

An operator, $\hat{O}$, is said to have the important property of being *Hermitian* for some set of functions if for every pair of functions, $\psi_i$ and $\psi_j$, from that set the following equivalence exists.

$$\int \psi_i^* \hat{O} \psi_j \, d\tau = \int \psi_j [\hat{O}^* \psi_i^*]\, d\tau \qquad (4\text{-}1a)$$

The difference between the left and right integrals is simply that instead of the operator being applied to $\psi_j$, the complex conjugate of the operator is applied to the other function, $\psi_i^*$. Eq. (4-1a) is also written as

$$<\psi_i \mid \hat{O} \psi_j> = <\hat{O} \psi_i \mid \psi_j> \qquad (4\text{-}1b)$$

in Dirac notation.

There are several useful theorems about Hermitian operators that should be established at this point.

***Theorem 4.1*** The expectation value of a Hermitian operator is a real number.

Given a wavefunction $\psi$ and an operator $\hat{G}$ that is Hermitian, then

| | |
|---|---|
| $<G> = <\psi \mid \hat{G} \psi>$ | by the definition of an expectation value |
| $= <\hat{G} \psi \mid \psi>$ | since $\hat{G}$ is Hermitian |

† Sometimes a second vertical bar is placed to the right of the operator, thereby sandwiching it between two bars, e.g., $<\psi \mid \hat{O} \mid \psi>$, with no change in meaning.

| | |
|---|---|
| $< G >^* = <\psi \mid \hat{G}\psi>^*$ | by taking the complex conjugate of the equation |
| $= <\hat{G}^*\psi^* \mid \psi^*>$ | the complex conjugate of an integral of a product of functions is the integral of the complex conjugates of the functions |
| $= <\psi \mid \hat{G}^{**}\ \psi^{**}>$ | from rewriting the integral, i.e., $<f \mid g> = <g^* \mid f^*>$ or $\int f^*(x)\, g(x)\, dx = \int g(x)\, f^*(x)\, dx$ |
| $= <\psi \mid \hat{G}\psi>$ | since $x = x^{**}$ |
| $= < G >$ | by the definition of an expectation value |

The result is that the expectation value of the operator is equal to its own complex conjugate. This can only be true for real numbers, since they have no imaginary part. So, the expectation value of a Hermitian operator must be real. Since observables or physical quantities that may be measured should have nonimaginary expectation values, then it must be that the operators associated with such quantities are Hermitian. The Hamiltonian, for instance, is a Hermitian operator.

***Theorem 4.2*** Eigenvalues of a Hermitian operator are real numbers.

Given a normalized wavefunction $\psi$ that is an eigenfunction of a Hermitian operator $\hat{G}$ with eigenvalue g, then

| | |
|---|---|
| $\hat{G}\psi = g\psi$ | given |
| $\psi^*\hat{G}\psi = \psi^* g\psi = g\ \psi^*\psi$ | multiplying the equation by $\psi^*$ |
| $<\psi \mid \hat{G}\psi> = g<\psi \mid \psi>$ | from integrating both sides of the equation |
| $= g$ | since $\psi$ is normalized |

Therefore, the eigenvalue is the same as the expectation value, which must be real by the previous theorem.

Two functions are said to be *orthogonal* over some region of space if there is a zero value to the integral over that region of the product of one function with the complex conjugate of the other function. That is, $<\psi_i|\psi_j> = 0$ is a statement of *orthogonality;* $\psi_i$ and $\psi_j$ are said to be orthogonal if this condition is satisfied. An analogous use of the term is that two geometrical vectors are orthogonal if their dot product is zero. The angle between two such vectors would be 90°, and in a like, though abstract, sense orthogonal functions may be thought of as being in completely different "directions" in a space of "functions."

*Theorem 4.3* Eigenfunctions of a Hermitian operator with different associated eigenvalues are orthogonal functions.

Given a set of normalized wavefunctions $\{\psi_i\}$ that are each eigenfunctions of a Hermitian operator $\hat{G}$ with associated eigenvalues $g_i$, none of which are equal, then

$<\psi_i|\hat{G}\psi_j> \;=\; <\psi_i|g_j\psi_j> \;=\; g_j<\psi_i|\psi_j>$ from given information

$<\psi_i|\hat{G}\psi_j> \;=\; <\hat{G}\psi_i|\psi_j>$ since $\hat{G}$ is Hermitian

$= \;<g_i\psi_i|\psi_j> \;=\; g_i<\psi_i|\psi_j>$ from given information

Since $g_i \neq g_j$, then $<\psi_i|\psi_j> \;=\; 0$.

This theorem shows that every eigenfunction of a Hermitian operator is orthogonal to every other eigenfunction with a different eigenvalue. Furthermore, it is possible to prove the following.

*Corollary 4.3.1* Eigenfunctions of a Hermitian operator with the same eigenvalues may be transformed into orthogonal functions while remaining eigenfunctions of the operator.

Transformed means combined linearly with each other in some way.

Since it has been stated that the Hamiltonian operator must be Hermitian, then this corollary theorem provides for all the solutions of the Schrödinger equation to be orthogonal functions. Of course, each quantum state of a system is a distinct solution of the Schrödinger equation, and so each state must be different from every other state in some way so as to be properly orthogonal.

## 4.2 SIMULTANEOUS EIGENFUNCTIONS

It is possible that there may be a set of functions that are simultaneously eigenfunctions of two different operators; however, if this happens, it means the operators commute.

***Theorem 4.4*** If a set of functions $\{f_i\}$ are eigenfunctions of two different operators, $\hat{A}$ and $\hat{B}$, then the operators commute.

$\hat{A} f_i = a_i f_i$ and $\hat{B} f_i = b_i f_i$ given

$\hat{A}\hat{B} f_i = \hat{A}(b_i f_i) = b_i \hat{A} f_i = b_i a_i f_i$ applying the operators

$\hat{B}\hat{A} f_i = \hat{B}(a_i f_i) = a_i \hat{B} f_i = a_i b_i f_i$ applying the operators

$\hat{A}\hat{B} f_i - \hat{B}\hat{A} f_i = a_i b_i f_i - b_i a_i f_i$ by subtraction

$= 0$ since the eigenvalues are numbers

Therefore, $\hat{A}\hat{B} - \hat{B}\hat{A} = 0$.

An important corollary, which will not be proved here, is

***Corollary 4.4.1*** If two operators commute, it is possible to find a set of functions that are simultaneously eigenfunctions of both.

When a wavefunction is an eigenfunction of an operator corresponding to some observable, the quantum mechanical uncertainty for a measurement of that observable is zero. Each and every measurement should yield the eigenvalue and nothing else. A consequence of the theorem about simultaneous eigenfunctions is that it may be possible to measure an observable of some system with no uncertainty provided that the corresponding operator commutes with the Hamiltonian.

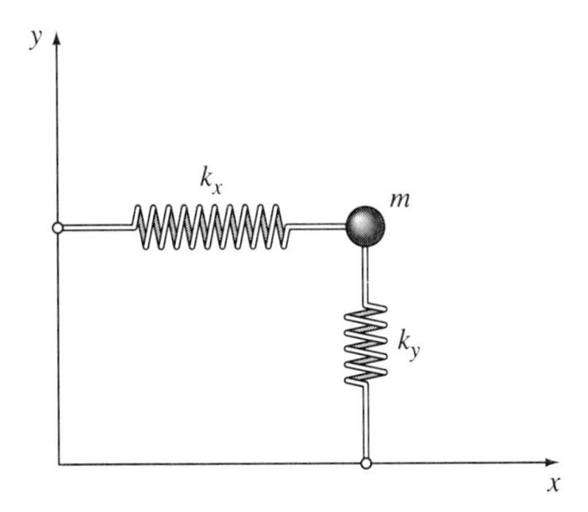


**FIGURE 4.1** A two-dimensional oscillator where a particle of mass m is attached to two harmonic springs with force constants $k_x$ and $k_y$. Each spring is contacted to an unmovable rod, but the connection points may slide freely along the rods (i.e., along the x-axis or along the y-axis). Thus, the potential energy for the $k_x$ spring depends only on the x-coordinate of the particle, while the potential energy of the $k_y$ spring depends only on the y-coordinate of the particle. The equilibrium lengths of the springs are taken to be zero for this example.

## 4.3 MULTIDIMENSIONAL PROBLEMS AND DEGENERACY

An example of a multidimensional quantum mechanical problem is the two dimensional oscillator shown in Fig. 4.1. The Hamiltonian for this problem is

$$\hat{H} = \frac{\hat{p}_x^2}{2m} + \frac{k_x x^2}{2} + \frac{\hat{p}_y^2}{2m} + \frac{k_y y^2}{2}$$

This is a separable Hamiltonian, and it has been written so as to make that apparent. The first two terms involve only the x-coordinate and the last two terms involve only the y-coordinate. As was done for separable, classical Hamiltonians, the two sets of terms may be freely designated as independent Hamiltonians:

$$\hat{H}_x \equiv \frac{\hat{p}_x^2}{2m} + \frac{k_x x^2}{2}$$

$$\hat{H}_y \equiv \frac{\hat{p}_y^2}{2m} + \frac{k_y y^2}{2}$$

$$\hat{H} = \hat{H}_x + \hat{H}_y$$

This merely highlights the fact that this particular Hamiltonian is separable.

The wavefunction for this system must be a function of the x- and y-position coordinates, $\psi = \psi(x,y)$. The Schrödinger equation,

$$(\hat{H}_x + \hat{H}_y)\,\psi(x,y) = E\,\psi(x,y)$$

is a separable differential equation. Separability is demonstrated by assuming that the wavefunction, $\psi$, takes on a product form; that is, the wavefunction is assumed to be a product of independent functions of x- and y-coordinates. These independent functions will be designated $X(x)$ and $Y(y)$, and so, the assumption is made that

$$\psi(x,y) = X(x)\,Y(y)$$

Now, if the Hamiltonian is applied to this assumed wavefunction, the component Hamiltonians act only on the respective coordinate functions.

$$\hat{H}_x X(x)\,Y(y) = Y(y)\,\hat{H}_x X(x)$$

$$\hat{H}_y X(x)\,Y(y) = X(x)\,\hat{H}_y Y(y)$$

Thus,

$$Y(y)\,\hat{H}_x X(x) + X(x)\,\hat{H}_y Y(y) = E\,X(x)\,Y(y)$$

and upon multiplying through by $1/(XY)$,

$$\frac{\hat{H}_x X(x)}{X(x)} + \frac{\hat{H}_y Y(y)}{Y(y)} = E \qquad (4\text{-}2)$$

This has the same form as the equation $f(x) + g(y) = c$. It is a sum of a function of the x-coordinate (only) and a function of the y-coordinate (only), and that sum is equal to a constant. Since x and y are independent variables, the only way for this to be true is if the function of x is constant and the function of y is also constant. Therefore, upon naming the two constants $E_x$ and $E_y$, we find that Eq. (4-2) implies the following.

$$\frac{\hat{H}_x X(x)}{X(x)} = E_x \quad \Rightarrow \quad \hat{H}_x X(x) = E_x X(x) \tag{4-3a}$$

$$\frac{\hat{H}_y Y(y)}{Y(y)} = E_y \quad \Rightarrow \quad \hat{H}_y Y(y) = E_y Y(y) \tag{4-3b}$$

$$\text{where} \quad E_x + E_y = E$$

There are now two Schrödinger equations. Recall that in a classical analysis of a mechanically separable problem, separate equations of motion are obtained too.

The procedure for working with the Schrödinger equation is general for all multidimensional problems that are separable. Separability implies two things:

i. The wavefunction is a product of independent coordinate functions of the separable variables.

ii. The total energy eigenvalue is a sum of energy terms associated with each separable coordinate.

If separability in a problem is not immediately apparent, a product form of the wavefunction may be tested just the same. If application of the Hamiltonian to the product form [e.g., $X(x)Y(y)$] yields an equation of the same type as Eq. (4-2), then the problem is separable. If the problem is not separable, the Schrödinger equation ends up as a coupled differential equation, and solution is usually more difficult.

Separability is as advantageous in quantum mechanics as in classical mechanics. For instance, the Schrödinger equations in Eq. (4-3) may be recognized as simple, one-dimensional, harmonic oscillator Schrödinger equations. Thus, the harmonic oscillator solutions given in Chap. 3 may be used directly. First, though, the independence of the x- and y-motions, which made for the separability, must be taken into account. In particular, the intrinsic vibrational frequencies of these one-dimensional oscillators are not necessarily the same since they depend on the respective spring constants. That is, there are two frequencies for this system, and they will be named $\omega_x$ and $\omega_y$.

$$\omega_x = \sqrt{\frac{k_x}{m}} \quad \text{and} \quad \omega_y = \sqrt{\frac{k_y}{m}} \tag{4-4}$$

Furthermore, there must be two independent quantum numbers since the extent of vibrational motion in the x-direction is independent of that in the y-direction. We may name these quantum numbers $n_x$ and $n_y$.

From the one-dimensional Schrödinger equation solutions, the energy eigenvalues are

$$E_x = (n_x + \frac{1}{2})\hbar\omega_x \quad \text{and} \quad E_y = (n_y + \frac{1}{2})\hbar\omega_y$$

On this basis, the energy of the system as a whole depends on both of the quantum numbers, since it is the sum of the x- and y-motion energies. It is appropriate to subscript the total energy, E, with the quantum numbers, and then the following may be said about the energy levels of the system.

$$E_{n_x n_y} = (n_x + \frac{1}{2})\hbar\omega_x + (n_y + \frac{1}{2})\hbar\omega_y \tag{4-5}$$

The quantum numbers, being those of a harmonic oscillator problem, may be any positive integer or zero.

The wavefunction, $\psi$, for the system as a whole is a product, and it must be labelled by the two quantum numbers in order to distinguish the different Schrödinger equation solutions.

$$\psi_{n_x n_y}(x,y) = X_{n_x}(x)\, Y_{n_y}(y) \tag{4-6}$$

The functions $X_{n_x}$ and $Y_{n_y}$ are simply one-dimensional harmonic oscillator wavefunctions. These product functions describe the *states* of the system as a whole.

An interesting situation would arise if this two-dimensional oscillator were *isotropic*, that is, if it were the same in all directions, or specifically if $k_x = k_y$. In that case, $\omega_x = \omega_y$, and the energy expression, Eq. (4-5), would become

$$E_{n_x n_y} = (n_x + n_y + 1)\,\hbar\,\omega \qquad \text{with } \omega = \omega_x = \omega_y$$

From this expression, the energies of various states may be tabulated by going through the sequence of all the allowed values of the two quantum numbers:

| $n_x$ | $n_y$ | $E_{n_x n_y} / \hbar\omega$ |
|---|---|---|
| 0 | 0 | 1 |
| 1 | 0 | 2 |
| 0 | 1 | 2 |
| 1 | 1 | 3 |
| 2 | 0 | 3 |
| 0 | 2 | 3 |
| 2 | 1 | 4 |
| 1 | 2 | 4 |
| 3 | 0 | 4 |
| 0 | 3 | 4 |

Notice that there are states, such as those with quantum numbers (1,0) and (0,1), that have exactly the same energy. This is termed *degeneracy*; states with the same energy are said to be degenerate with one another. The lowest energy state is the *ground state*, and in this case it is non-degenerate. It happens to be the only nondegenerate state for this problem. Also notice that at each higher energy level in this problem, the degeneracy, meaning the number of degenerate states, increases. Quantum mechanical degeneracy often has interesting manifestations in molecular spectroscopy, and examples will be given later.

## 4.4 VARIATION THEORY

For most chemical problems, the Schrödinger equation is not strictly separable and the differential equation is not easily solved by analytical means. Or more to the point, the problems are not simple or ideal. The techniques that are best used to find wavefunctions for complicated problems often turn out to be indirect, or at least they appear so. The techniques may also involve approximation as well. Variation theory and perturbation theory are the most powerful techniques and have been very important for understanding many quantum chemical systems.

The *variational principle* is the basis for the variational determination of a wavefunction. This principle tells us that for a given operator the eigenvalue of an eigenfunction is at a minimum value with respect to small adjustments that might be made to the function. As a consequence, the expectation value for some arbitrary function and the operator cannot be less than that minimum. In other words, for an arbitrary function, the expectation value is greater than or equal to the lowest eigenvalue of that operator. Now, the Schrödinger equation calls for finding eigenfunctions of a very specific operator, the Hamiltonian. So, the variational principle helps by saying that the expectation value of the Hamiltonian for any wavefunction we may guess will be greater than or equal to the true ground state energy, the lowest eigenvalue of the Hamiltonian. The expectation value of a guess wavefunction amounts to a guess of a system's energy eigenvalue, and the variation principle indicates that such a guess will err on the high side. This has an important consequence: A guess wavefunction can be improved – made more like the true Hamiltonian's eigenfunction – through any adjustment that lowers the expectation value of the energy. This is because the energy cannot be lowered "too much"; the variational principle dictates that the expectation value will never be less than the lowest eigenvalue.

For the variational principle to hold, the guess wavefunction must satisfy certain conditions that the true wavefunction satisfies. First, the function and its first derivative must be continuous in all spatial variables of the system and over the entire range of those variables. This requirement says the function does not ever stop and start up somewhere else and that it changes smoothly. Second, the function must be single-valued, which means that at any point in space, the function has only one unique value. Third, if the square of the function is integrated over all space, the result must be a finite value. These conditions follow from the postulates and specifically from the interpretation of the square of a wavefunction as a probability density. The first condition ensures that the integral of the probability density over any particular region of space is well-defined. The second condition ensures that the probability density at every point is unique. The third condition ensures that the function may be normalized. True wavefunctions, guess wavefunctions, and approximate wavefunctions must satisfy these conditions.

The process of adjusting guess wavefunctions, or trial wavefunctions, to minimize the expectation value of the energy is the *variational method*. A simple application is to the harmonic oscillator problem. The Hamiltonian is that of Eq. (3-7). The guess wavefunction, $\Gamma$, will be taken to be a Gaussian function with a parameter, $\alpha$, that will be adjustable.

$$\Gamma = \left(\frac{\alpha^2}{\pi}\right)^{1/4} e^{-(\alpha x)^2/2}$$

The constant in front of the exponential makes $\Gamma$ normalized for any choice of $\alpha$. Other functional forms might be used for the guess wavefunction; the Gaussian form is the example at hand. The Gaussian function satisfies the conditions mentioned above: It is continuous, has a continuous first derivative, is single-valued, and yields a finite value if it is squared and integrated from $-\infty$ to $\infty$.

The expectation value, designated W, is obtained by integration with the trial function.

$$\begin{aligned} W &= \langle \Gamma \mid \hat{H}\, \Gamma \rangle \\ &= \langle \Gamma \mid -\frac{\hbar^2}{2m}\frac{d^2}{dx^2} + \frac{1}{2}kx^2 \mid \Gamma \rangle \\ &= \frac{\hbar^2\alpha^2}{2m} + \left(\frac{k}{4} - \frac{\hbar^2\alpha^4}{4m}\right)\frac{1}{\alpha^2} = \frac{\hbar^2\alpha^2}{4m} + \frac{k}{4\alpha^2} \end{aligned}$$

The problem is to adjust $\alpha$ so as to minimize W. W is implicitly a function of the parameter $\alpha$, and so minimization may be accomplished by taking the first derivative.

$$\frac{dW}{d\alpha} = \frac{\hbar^2\alpha}{2m} - \frac{k}{2\alpha^3}$$

At any point where the first derivative of a function of one variable is zero, the value of the function itself is either a minimum or a maximum. [When the first derivative of f(x) is zero, f(x) is just starting to turn upward or else to turn downward, meaning it has reached a minimum or a maximum. From the second derivative, it can be determined if the point is a minimum or maximum.] For $W(\alpha)$, let that point be designated $\alpha_{min}$.

$$\left.\frac{dW}{d\alpha}\right|_{\alpha=\alpha_{min}} = 0 \qquad \Rightarrow \qquad \frac{\hbar^2 \alpha_{min}}{2m} - \frac{k}{2\,\alpha^3_{min}} = 0$$

The expression for the first-derivative function is set equal to zero when the variable is equal to that of the minimum (or maximum) point. That produces an equation for $\alpha_{min}$, and solving it yields

$$\alpha^2_{min} = \sqrt{km}\,/\,\hbar$$

This represents the best possible choice for adjusting $\alpha$ to improve the trial wavefunction, since the expectation value of the energy cannot be made any less with any other choice of $\alpha$. Comparison of this result with Eq. (3-10) shows that $\alpha_{min} = \beta$, where $\beta$ is the constant used in the harmonic oscillator eigenfunctions. (We might have expected that would work out.)

The expectation value of the energy is obtained by using the expression for $\alpha_{min}$ in $W(\alpha)$.

$$W(\alpha_{min}) = \frac{\hbar^2 \alpha^2_{min}}{4m} + \frac{k}{4\alpha^2_{min}}$$

$$= \frac{\hbar}{4}\frac{\sqrt{km}}{m} + \frac{\hbar}{4}\frac{k}{\sqrt{km}} = \frac{\hbar}{2}\sqrt{\frac{k}{m}}$$

This, as we know, is the true ground state energy of the harmonic oscillator as worked out in Chap. 3. Thus, by guessing the form of the harmonic oscillator wavefunction to be a Gaussian function, the variational method has determined *which* Gaussian has the lowest expectation value for the energy. And since the true ground state wavefunction is a Gaussian, this best Gaussian *is* the exact wavefunction. The power of this approach is evident upon realizing that the exact wavefunction and eigenenergy were obtained by relatively simple mathematics; there was no formal differential equation solving.

The variational method will not necessarily lead to the exact wavefunction, even though it happened to do so in the prior example.

The difficulty is that it is not generally possible to guess the exact form of a wavefunction. The variational method yields the best of whatever functional form is chosen, and that form may be only an approximation of the true wavefunction. This leads to the idea of increasing the degree of mathematical flexibility in the trial wavefunction, and that can be accomplished by using more parameters. For instance, in the harmonic oscillator problem, a more flexible trial function is

$$\Phi = N e^{-\alpha x^{\gamma}}$$

With $\alpha$ and $\gamma$ being adjustable, $\Phi$ could be a Gaussian (when $\gamma = 2$) just as well as something else ($\gamma \neq 2$). The flexibility has been accomplished by using two parameters. As a result, the expectation value of the energy of $\Phi$ is a function of more than one variable (parameter), and the minimization must be carried out with respect to both. Judicious choice of functional forms along with embedding many adjustable parameters is the key to applying variation theory to difficult problems. Using advanced computing systems, calculations have been reported in the research literature wherein wavefunctions for molecules have been found variationally with more than 10 million adjustable parameters!

To understand the variational principle itself, consider a problem with a known set of normalized Hamiltonian eigenfunctions, $\{\psi_i\}$, with corresponding eigenenergies, $\{E_i\}$. It can be proved that any function in the geometrical space of the problem can be represented as a superposition or linear combination of a set of Hamiltonian eigenfunctions. This is much like the idea of a Fourier expansion where a function can be represented by a sum of sine and/or cosine functions. In this sense, an arbitrary function *is* an arbitrary linear combination of the eigenfunctions. That is, anything that may serve as a guess function is the same thing as some linear combination of the $\psi_i$'s. If the coefficients, $\{a_i\}$, in this linear combination are considered arbitrary, then the guess function, $\Gamma$, is an arbitrary function.

$$\Gamma = N \sum a_i \psi_i$$

So, $\Gamma$ stands for any guess wavefunction whatsoever. N is the usual normalization constant. The expectation value of $\Gamma$ with the Hamiltonian is easily simplified because the $\psi$ functions are eigenfunctions of the Hamiltonian, and therefore are orthogonal.

$$W = \langle \Gamma \mid \hat{H} \Gamma \rangle$$

$$= N^2 \langle \sum_i a_i \psi_i \mid \hat{H} \sum_j a_j \psi_j \rangle$$

$$= N^2 \sum_i \sum_j a_i^* a_j \langle \psi_i \mid \hat{H} \psi_j \rangle$$

$$= N^2 \sum_i \sum_j a_i^* a_j E_j \langle \psi_i \mid \psi_j \rangle$$

$$= N^2 \sum_i \sum_j a_i^* a_j E_j \delta_{ij} = \sum_j (N^2 a_j^2) E_j \equiv \sum_j a_j'^2 E_j$$

The normalization factor, which is

$$N^2 = \left( \sum_i a_i^2 \right)^{-1}$$

has been absorbed into the primed coefficients in the last line. The primed coefficients squared must each be less than or equal to one because of the following.

$$\langle \Gamma \mid \Gamma \rangle = 1 = N^2 \sum_i \sum_j a_i^* a_j \langle \psi_i \mid \psi_j \rangle$$

$$= N^2 \sum_i a_i^2 = \sum_i a_i'^2$$

The sum of their squares is exactly 1.0, since $\Gamma$ is normalized, and so it would be impossible for any one coefficient to be greater than one.

The expectation value W must be greater than or equal to the lowest eigenenergy from the set $\{E_i\}$ because W is a sum of the eigenenergies, each multiplied by a number (e.g., $a_i'^2$) that is less than one. To illustrate this reasoning, consider the eigenenergies of the harmonic oscillator, $(n + 1/2)\hbar\omega$, and a guess wavefunction that is a linear combination of the harmonic oscillator functions, with the normalized expansion coefficient for the $n^{th}$ function being $a_n'$. The expectation value of the energy is

$$W = \frac{\hbar\omega}{2}\left( a_0'^2 + 3 a_1'^2 + 5 a_2'^2 + 7 a_3'^2 + \ldots \right)$$

and the condition on the coefficients is

$$a_0'^2 + a_1'^2 + a_2'^2 + \ \ldots \ = 1$$

By inspection, the smallest value for W is $\hbar\omega/2$, which is what we know to be the ground state energy. This will be obtained for the value of W only when $a_0'^2 = 1$ and the other coefficients are zero. Consider that relative to this choice, any change consistent with the constraints will make W a greater number. For instance, with $a_1'^2 = 0.9$, $a_2'^2 = 0.1$, and the other coefficients zero, $W = 3.2\ \hbar\omega$. This is the variational principle in action: The expectation value of the energy for an arbitrary function cannot be less than the lowest eigenvalue.

## 4.5 PERTURBATION THEORY

Perturbation theory offers another method for finding quantum mechanical wavefunctions. It is especially suited to problems that are similar to model or ideal situations differing in only some small way, which is the perturbation. For example, the potential for an oscillator might be harmonic except for a feature such as the small "bump" depicted in Fig. 4.2. Because the bump is a small feature, one expects the system's behavior to be quite similar to that of a harmonic oscillator. Perturbation theory would be a way to correct a description of the system, obtained from treating it as a harmonic oscillator, so as to account for the effects of the bump in the potential. Perturbation theory can yield exact wavefunctions and eigenenergies, but it can also be employed in an approximate way with much savings in effort.

In perturbation theory, the Hamiltonian for any problem is partitioned into two or more pieces. The first piece is one for which the eigenfunctions and eigenenergies are known, while everything else represents the perturbation. This first piece and the associated eigenfunctions and eigenenergies are distinguished in notation by a zero

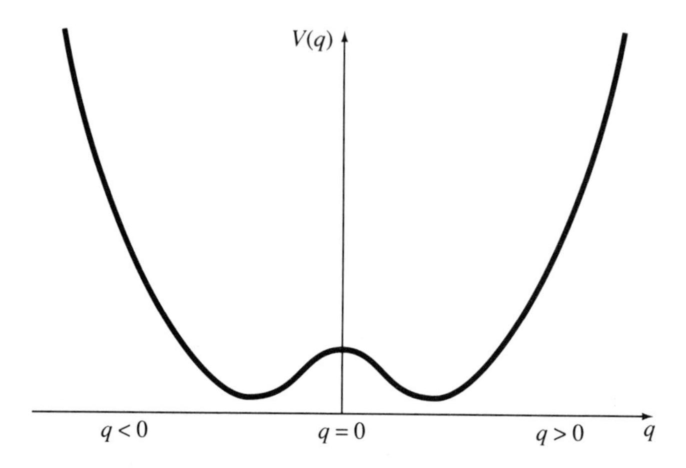


**FIGURE 4.2** A harmonic oscillator potential, a parabola, with a small perturbing potential, a bump in the middle. The composite potential is called a double-minimum potential or a *double-well potential.* An example of this type of potential is found for the water molecule. Water is a bent molecule, which means that the minimum of its potential energy as a function of the atomic positions is at a structure that is bent. We may envision a water molecule being straightened into a linear arrangement of the atoms, but this is necessarily a higher energy arrangement. If we continue, the molecule will be bent again but in the opposite sense:

$$\mathrm{H}\backslash_{\mathrm{O}}/\mathrm{H} \rightarrow \mathrm{HOH} \rightarrow {}_{\mathrm{H}}/^{\mathrm{O}}\backslash_{\mathrm{H}}$$

Eventually it will look like a mirror image of the original equilibrium structure, and so it will then be at a potential minimum. If the bending angle is taken to be a coordinate, then the potential energy as a function of the angle will have the form of a double well potential.

---

superscript. Thus, at the outset, one has a problem where the following Schrödinger equation has been solved.

$$\hat{H}^{(0)} \psi_i^{(0)} = E_i^{(0)} \psi_i^{(0)} \tag{4-7}$$

This is the zero-order perturbation equation. The complete Hamiltonian for the problem may be written with the introduction of a parameter, $\lambda$, which multiplies or scales the remaining piece of the Hamiltonian.

$$\hat{H} = \hat{H}^{(0)} + \lambda \hat{H}^{(1)} \tag{4-8}$$

If $\lambda$ is set to one, this expression gives the true Hamiltonian for the problem. If $\lambda$ is set to zero, it gives the Hamiltonian for the zero-order or unperturbed situation; there is, then, no contribution from $\hat{H}^{(1)}$. Thus, "tuning" $\lambda$ from zero to one lets us go smoothly from the situation we understand [Eq. (4-7)] to the problem we seek to solve. With this embedded parameter, the Schrödinger equation with $\hat{H}$ is really a family of equations covering all the choices of $\lambda$ values. Perturbation theory is a means of dealing with the entire family of equations, even though in the end, one is interested only in the case where $\lambda=1$.

Eq. (4-8) is an expansion of the Hamiltonian operator in a power series in $\lambda$. In fact, a general development of perturbation theory would allow for the possibility that some way of choosing the parameter might give rise to pieces of the Hamiltonian that depend on $\lambda$ quadratically, and to higher powers.

$$\hat{H} = \hat{H}^{(0)} + \sum_{n=1} \lambda^n \hat{H}^{(n)} \tag{4-9}$$

$\hat{H}^{(n)}$ is the $n^{th}$ order Hamiltonian. Writing the Hamiltonian as a power series implies writing the entire Schrödinger equation as an expansion in $\lambda$.

$$\left( \sum_{i=0} \lambda^i \hat{H}^{(i)} \right) \left( \sum_{j=0} \lambda^j \psi_n^{(j)} \right) = \left( \sum_{k=0} \lambda^k E_n^{(k)} \right) \left( \sum_{m=0} \lambda^m \psi_n^{(m)} \right) \tag{4-10}$$

The subscripts on the wavefunctions and on the energies identify a particular state as in Eq. (4-7).

When a particular order of Hamiltonian in Eq. (4-10) is applied to a particular order of wavefunction, the result, $\hat{H}^{(i)} \psi_n^{(j)}$, occurs in the Schrödinger equation with a factor of $\lambda^{(i+j)}$. On the right hand side of Eq. (4-10), energies multiply wavefunctions, and these have factors of $\lambda^{(k+m)}$. Clearly, terms will be found on the left hand side and on the right hand side that enter the equation with a factor of $\lambda^0$, $\lambda^1$, $\lambda^2$, and so on. The idea of perturbation theory is to approach a solution that is valid for the entire family of equations that arise from allowing $\lambda$ to be anything from zero to one. This implies a mathematical constraint that Eq. (4-10) remain valid for *all* choices of $\lambda$. At first that might seem impossible since some terms have a linear dependence on the parameter, some a quadratic dependence, and so on. The difference between the case of $\lambda = 1$ and $\lambda = 1/2$ is that a term linear in $\lambda$ would be diminished by one-half for the latter case, but a quadratic term would be diminished by one-fourth. The various terms would enter with a different weighting for every different choice of $\lambda$. However, that does not preclude a general solution. Instead, it dictates that the terms for any particular power of $\lambda$ on the left hand side of Eq. (4-10) must equal those on the right hand side. Below is a list of the first few equations which then result. They are found by multiplying out Eq. (4-10) and collecting terms according to the power of the $\lambda$ factor.

$$\lambda^0 \text{ terms:} \quad \hat{H}^{(0)} \psi_n^{(0)} = E_n^{(0)} \psi_n^{(0)}$$

$$\lambda^1 \text{ terms:} \quad \hat{H}^{(1)} \psi_n^{(0)} + \hat{H}^{(0)} \psi_n^{(1)} = E_n^{(1)} \psi_n^{(0)} + E_n^{(0)} \psi_n^{(1)}$$

$$\lambda^2 \text{ terms:} \quad \hat{H}^{(2)} \psi_n^{(0)} + \hat{H}^{(1)} \psi_n^{(1)} + \hat{H}^{(0)} \psi_n^{(2)}$$

$$= E_n^{(2)} \psi_n^{(0)} + E_n^{(1)} \psi_n^{(1)} + E_n^{(0)} \psi_n^{(2)}$$

Instead of one equation, the original Schrödinger equation, there are now an infinite number of equations because the $\lambda$ power series expansion is infinite. This is reasonable, though, since the ultimate solution is one that applies to the whole family of Schrödinger equations (i.e., the infinite number of choices of $\lambda$).

The $\lambda^0$ equation is the same as Eq. (4-7), and its solutions are known because of how the Hamiltonian was partitioned in the first place. Inspection of the $\lambda^1$ equation shows that it involves zero-order elements in addition to the first-order elements. The unknowns in this equation are the first-order correction to the wavefunction, $\psi_n^{(1)}$, and the first-order correction to the energy, $E_n^{(1)}$. In the $\lambda^2$ equation, zero-order and first-order elements are used in addition to the second-order corrections. Because of this pattern, the process for solving these equations must proceed from the zeroth equation to the first-order equation to the second-order equation, and so on. In this way, at any order, all the required lower order corrections will already be known.

Though more and more terms are involved as the order of $\lambda$, or the perturbation order, goes up, there are common steps in working with the equations. At each and every order, the energy correction is obtained by multiplying that $\lambda$-order equation by $\psi_n^{(0)*}$ and integrating over all space. There is a simplification that removes the two terms with $\psi_n^{(\lambda)}$ upon carrying out this integration. Notice that on the left hand side, this term will be $\hat{H}^{(0)}\psi_n^{(\lambda)}$, while on the right hand side it will be $E_n^{(0)}\,\psi_n^{(\lambda)}$. These can be grouped together on the left hand side as,

$$(\hat{H}^{(0)} - E_n^{(0)})\,\psi_n^{(\lambda)}$$

When this is integrated with the complex conjugate of the zero-order wavefunction, the result is zero.

$$\begin{aligned} <\psi_n^{(0)}\,|(\hat{H}^{(0)} - E_n^{(0)})\,\psi_n^{(\lambda)}> &= <(\hat{H}^{(0)} - E_n^{(0)})\psi_n^{(0)}\,|\,\psi_n^{(\lambda)}> \\ &= <(E_n^{(0)}\psi_n^{(0)} - E_n^{(0)}\psi_n^{(0)}\,|\,\psi_n^{(\lambda)}> = 0 \end{aligned}$$

The Hamiltonian operator is Hermitian, and the operation of multiplication by a constant is Hermitian; and that made it possible to exchange where the operator was being applied. With this simplification, the first-order equation yields an expression for the first-order correction to the energy.

$$\begin{aligned} <\psi_n^{(0)}\,|\,\hat{H}^{(1)}\psi_n^{(0)}> &= <\psi_n^{(0)}\,|\,E_n^{(1)}\,\psi_n^{(0)}> \\ &= E_n^{(1)}<\psi_n^{(0)}\,|\,\psi_n^{(0)}> = E_n^{(1)} \end{aligned} \qquad (4\text{-}11)$$

This reveals that the first-order correction to the energy is merely the expectation value of the first-order perturbing Hamiltonian with the zero-order wavefunction. The second-order equation yields an expression for the second-order correction to the energy.

$$<\psi_n^{(0)}\,|\,\hat{H}^{(2)}\psi_n^{(0)}> + <\psi_n^{(0)}\,|\,\hat{H}^{(1)}\psi_n^{(1)}> =$$

$$<\psi_n^{(0)}\,|\,E_n^{(2)}\,\psi_n^{(0)}> + <\psi_n^{(0)}\,|\,E_n^{(1)}\,\psi_n^{(1)}>$$

$$E_n^{(2)} = <\psi_n^{(0)}\,|\,\hat{H}^{(2)}\,\psi_n^{(0)}> + <\psi_n^{(0)}\,|(\hat{H}^{(1)} - E_n^{(1)})\,\psi_n^{(1)}> \qquad (4\text{-}12)$$

Using this equation, though, requires knowing the first-order correction to the wavefunction.

To find the corrections to the wavefunction, an idea from the previous section is used that an arbitrary function in some quantum mechanical space can be expressed as a linear combination of eigenfunctions of any Hamiltonian for that space. In perturbation theory, the zero-order equation is presumed to have been solved, and so the prerequisite, a set of eigenfunctions must exist. Thus, the arbitrary functions, first- or second-order corrections to the wavefunction, and so on, can each be expressed as a linear combination of the zero-order functions. The coefficients in that linear expansion constitute the information that has to be obtained. Let us apply this to the first-order equation and let $c_i$ be the expansion coefficient for the $i^{th}$ zero-order function.

$$\psi_n^{(1)} \equiv \sum_i c_i \psi_i^{(0)} \tag{4-13}$$

If we group Hamiltonians and energies as was done above, then Eq. (4-13) may be used with the first-order Schrödinger equation to give

$$(\hat{H}^{(1)} - E_n^{(1)}) \psi_n^{(0)} = -(\hat{H}^{(0)} - E_n^{(0)}) \psi_n^{(1)}$$

$$= -(\hat{H}^{(0)} - E_n^{(0)}) \sum_i c_i \psi_i^{(0)}$$

$$= -\sum_i c_i (E_i^{(0)} - E_n^{(0)}) \psi_i^{(0)} \tag{4-14}$$

This result can be used to extract one of the desired coefficients by multiplying by the complex conjugate of one of the zero-order wavefunctions and integrating over all space. Let that one function be $\psi_j^{(0)}$, with $j \neq n$ for now.

$$< \psi_j^{(0)} \mid (\hat{H}^{(1)} - E_n^{(1)}) \psi_n^{(0)} > = -\sum_i c_i (E_i^{(0)} - E_n^{(0)}) < \psi_j^{(0)} \mid \psi_i^{(0)} >$$

$$< \psi_j^{(0)} \mid \hat{H}^{(1)} \psi_n^{(0)} > - E_n^{(1)} < \psi_j^{(0)} \mid \psi_n^{(0)} > = -\sum_i c_i (E_i^{(0)} - E_n^{(0)}) \delta_{ij}$$

$$< \psi_j^{(0)} \mid \hat{H}^{(1)} \psi_n^{(0)} > \;=\; -c_j \, (E_j^{(0)} - E_n^{(0)})$$

$$\therefore \quad c_j \;=\; < \psi_j^{(0)} \mid \hat{H}^{(1)} \psi_n^{(0)} > / \; (E_n^{(0)} - E_j^{(0)}) \tag{4-15}$$

By letting j take on all values except n, we may use this equation to find all the coefficients except $c_n$. [We cannot use Eq. (4-15) with j = n because the denominator is zero then.]

The result presented in Eq. (4-15) can be discussed in several ways. We see that a particular $c_j$ will be zero if the integral quantity involving the zero-order wavefunctions for the $n^{th}$ and $j^{th}$ states with the perturbing Hamiltonian is zero. Thus, if there is no *coupling* or *interaction* between two states brought about by a perturbation, then there will be no *mixing* of their wavefunctions to first order. Also, if the difference in zero-order energies is large, the extent of mixing will be small because this difference is in the numerator in Eq. (4-15).

The procedure that has been developed for first-order corrections to the wavefunction can be used at second and higher orders as well. The resulting expressions will be more complicated. The unanswered question at this point is the value of the coefficient $c_n$. This comes from another procedure, and it too can be used for all orders. This procedure develops from the normalization condition, and specifically from requiring that the perturbed wavefunction be normalized for any choice of the perturbation parameter $\lambda$.

$$1 \;=\; < \psi_n \mid \psi_n > \;=\; < \psi_n^{(0)} + \lambda \, \psi_n^{(1)} + \ldots \mid \psi_n^{(0)} + \lambda \, \psi_n^{(1)} + \ldots >$$

$$= \; < \psi_n^{(0)} \mid \psi_n^{(0)} > \; + \; \lambda \left( < \psi_n^{(1)} \mid \psi_n^{(0)} > \; + \; < \psi_n^{(0)} \mid \psi_n^{(1)} > \right)$$

$$+ \; \ldots$$

From the power series expansion, the normalization condition is now written as a sum of terms that depend on different orders of $\lambda$. The first right hand side term is, of course, equal to one, and so all the remaining terms must add up to zero. Such a result will be true for any choice of $\lambda$ only if each term is itself zero. The term that has $\lambda$ to the first power consists of an integral with its complex conjugate. If that integral is zero, then so is its complex conjugate. Therefore,

$$0 = \langle \psi_n^{(0)} | \psi_n^{(1)} \rangle = \langle \psi_n^{(0)} | \sum_i c_i \psi_i^{(0)} \rangle$$

$$= \sum_i c_i \langle \psi_n^{(0)} | \psi_i^{(0)} \rangle = \sum_i c_i \delta_{ni} = c_n$$

This establishes that $c_n$ is zero. For higher order corrections to the wavefunction, it will not necessarily be that the $n^{th}$ coefficient is zero; this is a special result at first order.

The perturbation derivations can be carried out to any desired order, but the idea of perturbation theory is that the perturbation, whatever it is, makes a small correction to the zero-order picture. In the typical application, the lowest few orders provide nearly all of the correction needed to make the energies and wavefunctions exact. Of course, *nearly* all is not all; to stop at some low order of perturbation theory (i.e., to truncate the expansion in $\lambda$) is to make an approximation. The quality of the approximation that we make, or the order at which we truncate, is our choice when we use this approach.

An illustration of the efficacy of low-order perturbation theory, in fact, of the first order energy corrections, is given in Fig. 4.3. The model problem is that of a slightly altered harmonic vibrational potential, the type seen in Fig. 4.2. The result of an extensive variational treatment that closely approaches the exact energies and wavefunctions for the lowest several states is shown along with the energy levels obtained from first-order perturbation theory, taking the bump in the potential to be the perturbation of the otherwise harmonic system. The energies of the variational and perturbational treatments may be compared with the energies of the unperturbed (harmonic) oscillator. From that, we see that the first-order corrections are similar to the energy changes obtained from a variational treatment. Also, the correspondence between the variational energy and the first-order perturbation theory energy improves as one goes to higher energy levels. This is because for the higher energy states, the relative effect of the bump in the potential is diminished, and so, perturbation theory at low order is even more appropriate.

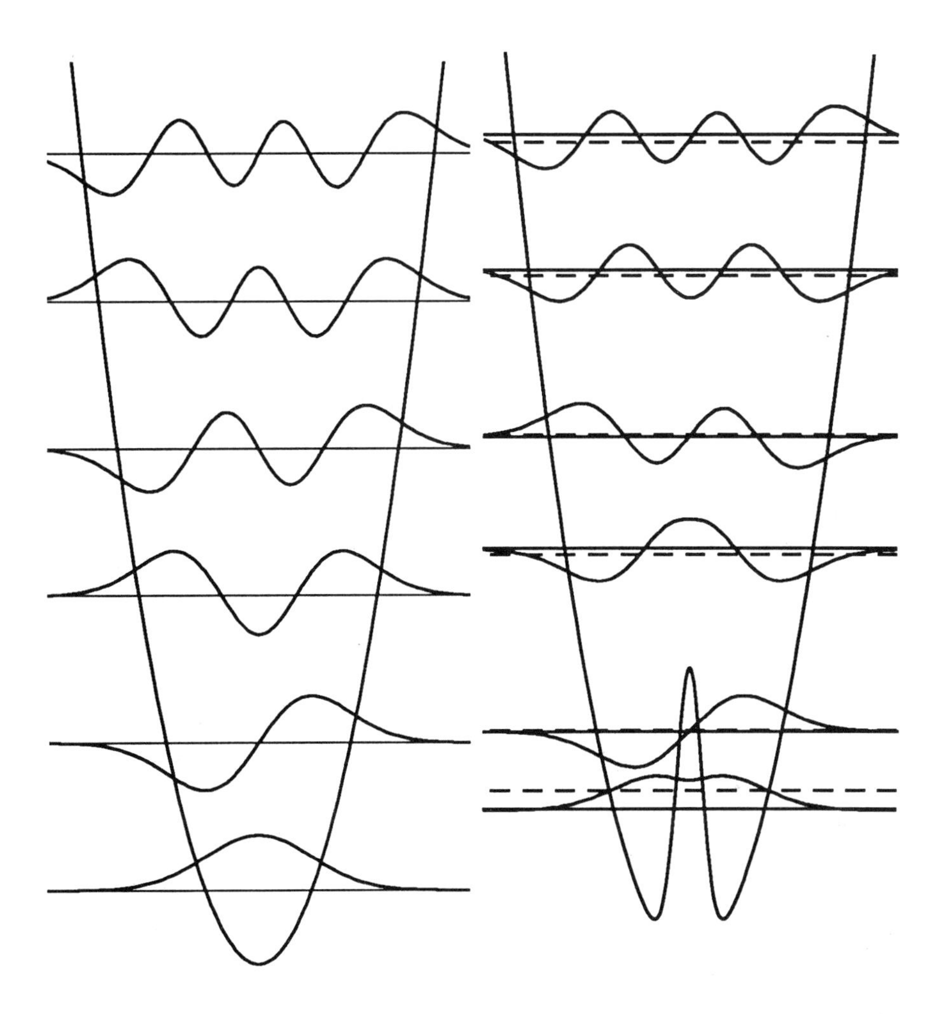

**FIGURE 4.3** On the left is a harmonic vibrational potential with the exact energy levels and corresponding wavefunctions drawn for the first several states. On the right is the same potential augmented, or perturbed, by a bump in the middle. The solid horizontal lines drawn with this potential are the energy levels obtained from an extensive variational treatment of this problem, whereas the dashed lines are the energy levels obtained from first-order perturbation theory, using the system on the left as the zero-order picture and the bump as the perturbation. The wavefunctions drawn for the perturbed system are those obtained from the variational treatment, and they demonstrate the detailed changes in the wavefunctions from the perturbing potential.

## 4.6 TIME DEPENDENCE AND TRANSITIONS

To this point, attention has been limited to wavefunctions that do not evolve in time. However, Hamiltonians may include time-dependent elements that cause the system to change, and an important case of this sort is the effect of electromagnetic radiation. The character of light is that of electric and magnetic fields that oscillate in space and time. When light impinges on a molecule, the oscillating fields may give rise to an interaction, i.e., an element of the complete Hamiltonian, and so the time dependence of the fields then enters the Hamiltonian. We shall see that the interaction with light can induce transitions among the states of a system, and the mechanisms for this are the basis for molecular spectroscopy experiments.

When time, t, becomes a variable in a quantum mechanical problem, an associated operator is needed. Like a position coordinate, time as an operator is just multiplication by t. Also like a position coordinate, time has a conjugate variable with its own associated operator. Now, two operators are conjugate if their commutator is $i\hbar$. This was the case for the position operator x and the momentum operator $p_x$. To find the operator that is conjugate with time, an operator equation is employed. Using f to designate an arbitrary function, and G to be the operator we wish to find, then,

$$i\hbar = [\hat{G},t] \quad \Rightarrow \quad \hat{G}\,t\,f - t\,\hat{G}\,f = i\hbar\,f$$

$$\Rightarrow \quad \hat{G} = i\hbar\frac{\partial}{\partial t}$$

Dimensional analysis of this operator shows that it will produce an energy quantity (namely, erg-sec/sec = erg) upon applying it to a wavefunction. This operator, which will be designated as $\hat{E}$ from now on, plays an important role in the time dependent wavefunctions. In the time dependent generalization of Postulate III (the Schrödinger equation), this operator replaces E, the energy eigenvalue. The *time dependent Schrödinger equation* (TDSE) is

$$\hat{H}\Psi = \hat{E}\Psi \tag{4-16}$$

$\Psi$ is an explicit function of time in this equation, and the Hamiltonian may have an explicit time dependence also. The generalization of Postulate III is that the solutions of the TDSE are the wavefunctions that describe a system in both space and time.

The time-dependent Schrödinger equation is a generalization of the time-independent Schrödinger equation (TISE) but does not invalidate the TISE. This is because the TDSE is always separable in space and time in those cases where the Hamiltonian has no explicit time dependence (which are the only kinds of cases considered until now). We may demonstrate this separability with the following steps, where only one spatial coordinate, x, is used for simplicity. The Hamiltonian operator is taken to have no explicit time dependence.

$$\hat{H}(x)\,\Psi(x,t) = i\hbar \frac{\partial}{\partial t}\Psi(x,t)$$ the TDSE

$$\Psi(x,t) = \psi(x)\,\phi(t)$$ trial product form of $\Psi$

$$\frac{\hat{H}(x)\,\psi(x)}{\psi(x)} = \frac{i\hbar}{\phi(t)}\frac{\partial\phi(t)}{\partial t}$$ substitution into TDSE and division by $\psi(x)\,\phi(t)$

$$\equiv E$$ the left and right hand sides are functions of independent variables and so must each equal a constant, E

$$\therefore \quad \hat{H}(x)\,\psi(x) = E\,\psi(x)$$ separated equation in x

$$\therefore \quad i\hbar\frac{\partial\phi(t)}{\partial t} = E\,\phi(t)$$ separated equation in t

Thus, with no explicit time dependence in the Hamiltonian, the TDSE is separable into a spatial differential equation and a differential equation in time. The differential equation in the spatial coordinate(s) is seen to be the familiar time-independent Schrödinger equation.

The general solution of the differential equation in time that was obtained for the time-independent Hamiltonian case is

$$\phi(t) = \exp(-iEt/\hbar) \qquad (4\text{-}17)$$

This function is sometimes referred to as a *phase* or said to give the phase of the wavefunction. If the TISE is solved for a particular problem, the result is a spatial wavefunction, but multiplying that function by the $\phi(t)$ in Eq. (4-17) gives a product function that is necessarily a solution of the TDSE. Such a product wavefunction is called a *stationary state* wavefunction because the probability density function is independent of time: $\phi^*\phi = 1$. That is, if $\Psi(x,t) = \psi(x)\,\phi(t)$, then $\Psi(x,t)^*\Psi(x,t) = \psi^*(x)\,\psi(x)$.

It is an important result that even for time-independent Hamiltonian problems, stationary state wavefunctions are not the only possible solutions of the TDSE. In fact, any arbitrary superposition (or linear combination) of different stationary state wavefunctions will be a solution of the TDSE. To illustrate this point, let $\hat{H}_o$ be some particular Hamiltonian with no explicit time dependence and let $\{\psi_i\}$ be the set of its eigenfunctions with associated eigenvalues $\{E_i\}$.

$$\hat{H}_o \psi_i = E_i \psi_i \tag{4-18}$$

The stationary state wavefunctions for this problem are

$$\Psi_i = \psi_i \, e^{-iE_i t/\hbar} \tag{4-19}$$

An arbitrary superposition of these states is a linear combination of these functions with unspecified or arbitrary coefficients.

$$\Gamma = \sum_i c_i \Psi_i$$

We can show that $\Gamma$ is, in fact, a solution of the TDSE by operating on it with $\hat{H}_o$ and with the energy operator and comparing the results:

$$\hat{H}_o \Gamma = \sum_i c_i \hat{H}_o \psi_i \, e^{-iE_i t/\hbar} = \sum_i c_i E_i \psi_i \, e^{-iE_i t/\hbar}$$

$$i\hbar \frac{\partial}{\partial t}\Gamma = \sum_i c_i \psi_i (i\hbar) \frac{\partial}{\partial t} e^{-iE_i t/\hbar} = \sum_i c_i E_i \psi_i \, e^{-iE_i t/\hbar}$$

The results are identical. So, even though $\Gamma$ is not an eigenfunction of the Hamiltonian, it is a solution of the TDSE.

When the Hamiltonian has explicit time dependence, separability of the TDSE into time and spatial differential equations is usually not possible. Analytical solution of the TDSE may be a horrible task. We will take a narrower view of time-dependent Hamiltonian problems, and we will treat the time dependence in the Hamiltonian as a perturbation of a system with a time-independent Hamiltonian. In other words, we will consider any complete Hamiltonian as consisting of two pieces, a time-dependent piece, H', and a time-independent piece, $H_o$; H' is a perturbation of the $H_o$ system.

As in the prior two sections, a suitable linear combination of the eigenfunctions or of Schrödinger equation solutions can form any valid wavefunction for the system. Use of this fact converts the task of solving the differential Schrödinger equation into a task of finding the linear expansion coefficients. The same thing is to be done now, except that the expansion coefficients must be allowed to be functions of time so that the TDSE wavefunction can evolve in time. The linear combination will be made from the stationary states of the TDSE involving just $H_o$ [Eq. (4-18)]. Thus, for the Schrödinger equation,

$$(\hat{H}_o + \hat{H}')\,\Phi \;=\; i\hbar \frac{\partial \Phi}{\partial t} \tag{4-20}$$

we will use the following general expansion for the wavefunction $\Phi$ where the stationary state wavefunctions, $\Psi_i$, are as given in Eq. (4-19).

$$\Phi \;=\; \sum_i a_i(t)\, \Psi_i \tag{4-21}$$

Fully solving Eq. (4-20), then, means finding all the $a_i(t)$ functions. The first step is to substitute Eq. (4-21) for $\Phi$ into Eq. (4-20). At the same time, it is helpful to write out the stationary state wavefunctions as the products of spatial and time functions via Eq. (4-19).

$$\sum_i a_i(t)\, e^{-iE_it/\hbar}\, \hat{H}_o \psi_i \;+\; \sum_i \hat{H}'\, a_i(t)\, e^{-iE_it/\hbar}\, \psi_i \;=\; i\hbar \sum_i \psi_i \frac{\partial}{\partial t} a_i(t)\, e^{-iE_it/\hbar}$$

Notice that on the left hand side of this expression, $\hat{H}_o\psi_i$ can be replaced with $E_i\psi_i$. On the right hand side, the partial differentiation will yield exactly the same thing from the differentiation of the exponential, plus an additional term. Thus, this expression simplifies to

$$\sum_i \hat{H}' a_i(t)\, \psi_i\, e^{-i E_i t/\hbar} = i\hbar \sum_i \dot{a}_i(t)\, \psi_i\, e^{-i E_i t/\hbar} \tag{4-22}$$

where $\dot{a}_i(t)$ means the time derivative of that function.

Since the stationary state wavefunctions used in the expansion for $\Phi$ are orthonormal, a useful thing happens if Eq. (4-22) is multiplied by any stationary state wavefunction, such as $\Psi_k$, and then integrated over all spatial coordinates. Orthonormality simplifies the right hand summation.

$$\sum_i < \Psi_k \mid \hat{H}' a_i(t)\, \Psi_i > \;=\; i\hbar \sum_i \dot{a}_i(t) < \Psi_k \mid \Psi_i >$$

$$= i\hbar\, \dot{a}_k(t) \tag{4-23}$$

Since k could be any index, Eq. (4-23) represents a whole set of coupled differential equations. If the integral quantities happen to be known, it might be possible to solve the set of equations and obtain the desired $a_i(t)$ functions. Short of doing that, we restrict attention to the short time behavior of $\Phi$ starting from a point, defined as $t = 0$, where the wavefunction $\Phi$ is one and only one stationary state, i.e., where for some state designated "initial,"

$$a_{initial}(0) = 1 \qquad \text{and} \qquad a_{i \neq initial}(0) = 0$$

Eq. (4-23) is valid for any time including the specific time $t = 0$. So, substituting in the $t = 0$ values for the a functions yields

$$< \Psi_k \mid \hat{H}'\, \Psi_{initial} > \;=\; i\hbar\, \dot{a}_k(0) \tag{4-24}$$

The significance of this result is that for some choices of k, the left hand side integral might be zero. For those choices, the first time derivative of the $a_k(t)$ function at $t = 0$ would have to be zero. Since $a_k(t)$ gives the weighting of the k stationary state in the $\Phi$ function, the zero value for the first derivative means that for short times at least, the weighting of the k stationary state will be unchanging. And by the conditions we chose for $t = 0$, this means that when k is any index other than "initial" the weighting will be zero. The system will not be evolving into the k stationary state (at least in the short time limit). So, whether the $\Phi$ function can evolve from some initial stationary state into another because of the effect of a time-dependent Hamiltonian H' all hinges on whether the integral in Eq. (4-24) is zero or not zero (at least in the short time limit).

When H' corresponds to the interaction with electromagnetic radiation in some particular experimental arrangement, then the qualitative distinction between a zero and a nonzero integral in Eq. (4-24) becomes the basis for spectroscopic *selection rules*. When the integral is nonzero, the *transition* from the initial state to the k state or "final" state is said to be *allowed*. When the integral is identically zero, the transition is said to be *forbidden* or not allowed. As we shall see in later chapters, allowed transitions lead to characteristic features of molecular spectra. Forbidden transitions are often not observed.

## 4.7 MATRIX METHODS FOR LINEAR VARIATION THEORY*

A special and powerful use of variation theory is with linear variational parameters. That means that the trial wavefunction is taken to be a linear combination of functions in some chosen set. The adjustable parameters are the expansion coefficients of each of these functions. This is, of course, a specialization of the way in which variation theory may be used, but it is powerful because the resulting equations take the form of matrix expressions. Solving the Schrödinger equation becomes a problem in linear algebra, and such problems are ideally suited to computer solution.

For any problem to be treated by linear variational methods, a *basis set* must be selected at the outset. This is simply some set of functions of the coordinates for the problem under study. In the best circumstances, the functions are chosen to be part of the basis set because they are close to the anticipated form of the true wavefunction. The number of functions is not restricted. A few may be used, or many may be used. The basis set may even have an infinite number, though the discussion here will assume a finite basis set.

The linear variational wavefunction, given as a *basis set expansion*, is a linear combination of the functions in the basis. With $\Gamma$ as the wavefunction and $\{\phi_i\}$ designating the set of basis functions, the expansion is

$$\Gamma = \sum_i^N c_i \phi_i \tag{4-25}$$

N is the number of functions in the basis set. The functions in the set are distinguished by the subscript i. The coefficients, $c_i$, are expansion coefficients; they are the adjustable parameters. Notice that basis set expansions have already been used: The first-order perturbation theory corrections to the wavefunction were obtained as an expansion in a basis of the zero-order functions.

The expectation value of the energy of $\Gamma$ is the quantity to be minimized. This is given by the following expression, subject to a normalization constraint on $\Gamma$.

$$\begin{aligned} \langle \Gamma | H \Gamma \rangle &= \langle \sum_i^N c_i \phi_i \,|\, H \sum_j^N c_j \phi_j \rangle \\ &= \sum_i^N \sum_j^N c_i^* \langle \phi_i | H \phi_j \rangle c_j \end{aligned} \tag{4-26}$$

This double-summation expression has a form that can be expressed with matrices (see Appendix I). It happens to be of the form of a row of values (coefficients) times a matrix of (integral) values times a column of coefficients, with the result being a scalar. The rank-one matrices of coefficients will be called column or row vectors, designated as (see Appendix I)

$$\vec{\mathbf{c}} = \begin{pmatrix} c_1 \\ c_2 \\ c_3 \\ \cdots \\ c_N \end{pmatrix} \quad \text{and} \quad \vec{\mathbf{c}}^T = (c_1 \ c_2 \ c_3 \ \cdots \ c_N)$$

It is also helpful to have a special symbol for the complex conjugate of the row vector: $\vec{c}^{\dagger} = \vec{c}^{T*}$. The integral values are arranged into a rank-two N-by-N matrix designated **H**.

$$\mathbf{H} = \begin{pmatrix} <\phi_1 | H\phi_1> & <\phi_1 | H\phi_2> & <\phi_1 | H\phi_3> & \ldots & <\phi_1 | H\phi_N> \\ <\phi_2 | H\phi_1> & <\phi_2 | H\phi_2> & <\phi_2 | H\phi_3> & \ldots & <\phi_2 | H\phi_N> \\ <\phi_3 | H\phi_1> & <\phi_3 | H\phi_2> & <\phi_3 | H\phi_3> & \ldots & <\phi_3 | H\phi_N> \\ & & \ldots & & \end{pmatrix}$$

**H** is called the *matrix representation* of the Hamiltonian operator. A matrix representation of an operator is a matrix of integral values arranged into rows and columns according to the basis functions. Clearly, the values in a matrix representation are dependent on the functions that were selected for the basis set. A different basis set implies a different matrix representation. Wherever it is important to keep track of the basis used in the representation, a superscript is added to the designation of the matrix, e.g., $\mathbf{H}^{\phi}$, and it identifies the particular function set. Now, the quantity in Eq. (4-26) can be written with the coefficient vectors and the Hamiltonian matrix in a very simple form.

$$<\Gamma | H\Gamma> = \vec{c}^{\dagger} \mathbf{H} \vec{c} = \sum_{i,j} c_i^* \mathbf{H}_{ij} c_j$$

Again, this is the form of an N-long row of numbers times an N-by-N square array times an N-long column of numbers, and that produces just one value (not a vector or matrix).

The overlap of the $\Gamma$ function with itself is the dot product of the coefficient vector and its complex conjugate transpose if, as we assume here, the basis functions are orthonormal.

$$<\Gamma | \Gamma> = \vec{c}^{\dagger} \vec{c} = \sum_j c_j^* c_j$$

If this is divided into $<\Gamma | H\Gamma>$, the resulting value is the energy expectation whatever the normalization of $\Gamma$. This is a way of imposing the normalization constraint.

$$W = \frac{<\Gamma | H\Gamma>}{<\Gamma | \Gamma>}$$

W is a function of the coefficients in $\vec{c}$. To apply variation theory, W must be minimized with respect to each of the elements of $\vec{c}$. This means taking the first derivative and setting that function to zero at the point where the coefficients lead to the minimum value of W.

$$\left.\frac{\partial W}{\partial c_i}\right|_{\vec{c}=\vec{c}_{min}} = 0 \qquad (4\text{-}27)$$

There will be one equation for each coefficient.

The partial differentiation of W with respect to one of the coefficients yields the following.

$$\frac{\partial W}{\partial c_i^*} = \frac{\sum_j \mathbf{H}_{ij} c_j}{<\Gamma | \Gamma>} - \frac{c_i}{<\Gamma | \Gamma>} \left\{ \frac{<\Gamma | H\Gamma>}{<\Gamma | \Gamma>} \right\}$$

(We may also differentiate† with respect to $c_i$ instead of with respect to $c_i^*$.) Notice that the quantity in brackets is W. So, when this first-derivative function is evaluated at the coefficient values that minimize W, i.e. $\vec{c}_{min}$, then this bracketed quantity must equal E, the variational energy. With this replacement for the bracketed quantity, we have that

$$\left.\frac{\partial W}{\partial c_i^*}\right|_{\vec{c}=\vec{c}_{min}} = \frac{1}{<\Gamma | \Gamma>} \left( \sum_j \mathbf{H}_{ij} c_{j\text{-}min} - c_{i\text{-}min} E \right)$$

Setting this to zero following the condition for minimization of W yields

$$\sum_j \mathbf{H}_{ij} c_{j\text{-}min} - c_{i\text{-}min} E = 0$$

† The differentiation of W, as shown, is with respect to a complex conjugate of one of the coefficients and is carried out by treating each coefficient as independent of its complex conjugate. By carrying out the differentiation with respect to complex conjugates, equations are developed for the simple coefficients. Should the coefficients be taken to be real, which would mean the coefficient and its complex conjugate are the same, then differentiation yields the same expression, but times 2. The factor of 2, though, makes no difference in the coupled equations because of setting the derivative to zero.

Clearly, there will be one such equation for each choice of the index i. The whole set of these equations can be arranged into a single matrix expression.

$$\mathbf{H}\,\vec{c}_{min} = \vec{c}_{min}\,E \tag{4-28}$$

This expression is an important and common expression called a matrix eigenvalue equation. It says that the matrix **H** operating on (or multiplying) the vector of coefficients that minimizes W is equal to that vector times a constant, which is the energy eigenvalue. This result means that in general applying linear variation theory will amount to finding an eigenvector of the matrix representation of the Hamiltonian.

There are numerous means for solving the matrix equation given in Eq. (4-28) . Regardless of which algorithm is applied, the process of finding the eigenvalues and the coefficient vectors is referred to as a *diagonalization* of **H** because it amounts to a transformation of **H** from the original basis into a basis where it is diagonal. Diagonalization is extremely well-suited to computers, and there are many well-crafted computer programs in use for diagonalizing matrices. Some are general routines, while others are specialized for the forms that certain matrices take on. For a matrix with N rows and N columns, it is usual for a diagonalization routine to require a number of multiplications of pairs of real numbers on the order of $N^3$. This means that the computational cost grows with the matrix dimension as $N^3$.

## 4.8 FIRST-ORDER DEGENERATE PERTURBATION THEORY ✯

Degeneracy complicates perturbation theory. If there is a state, $\psi_j$, that is degenerate with the state of interest, $\psi_n$, at zero-order, then the first-order correction to the wavefunction cannot be found with Eq. (4-15) since the energy denominator is zero: $E_n^{(0)} = E_j^{(0)}$. Of course there is way around this complication.

If a set of states, which may be a subset of the states of some system, happens to be degenerate with respect to some operator, then any normalized linear combination of those states will also be an eigenfunction of that operator with the same eigenvalue. For instance, consider a set of functions $\{\phi_i, i = 1, ..., N\}$ that are degenerate, normalized eigenfunctions of a zero-order Hamiltonian. That is,

$$\hat{H}^{(0)} \phi_i = E_n^{(0)} \phi_i$$

Any linear combination of these functions will be an eigenfunction of this Hamiltonian, as the following steps show.

$$\hat{H}^{(0)} \sum_{j=1}^{N} c_j \phi_j = \sum_{j=1}^{N} c_j \hat{H}^{(0)} \phi_j$$

$$= \sum_{j=1}^{N} c_j E_n^{(0)} \phi_j = E_n^{(0)} \sum_{j=1}^{N} c_j \phi_j$$

Consequently, the zero-order wavefunction in the perturbation treatment for the nth state cannot be presumed to be any particular one of the $\phi_i$ functions. Instead, it must be regarded as an undetermined linear combination of the functions in the degenerate subset.

$$\psi_n^{(0)} = \sum_{i=1}^{N} c_j \phi_j \qquad (4\text{-}29)$$

This is substituted into the first-order perturbation theory Schrödinger equation to give

$$\hat{H}^{(1)} \sum_{j=1}^{N} c_j \phi_j + \hat{H}^{(0)} \psi_n^{(1)} = E_n^{(1)} \sum_{j=1}^{N} c_j \phi_j + E_n^{(0)} \psi_n^{(1)} \qquad (4\text{-}30)$$

Now, solution of this equation must yield not only the first-order correction to the energy and the first-order correction to the wavefunction, but also the value of the expansion coefficients in the zero-order wavefunction, the $c_j$'s.

In nondegenerate perturbation theory, the procedure for obtaining the first-order correction to the energy was to multiply by the zero-order wavefunction for the state of interest and integrate. The result, Eq. (4-11), was that the first-order correction to the energy was the expectation value of the perturbing Hamiltonian with the zero-order wavefunction. In the case of degeneracy, it is necessary to carry out this process with each of the degenerate functions. So, we may take one function from the set and multiply Eq. (4-30) by it and then integrate.

$$< \phi_k \mid \hat{H}^{(1)} \sum_{j=1}^{N} c_j \phi_j > + < \phi_k \mid \hat{H}^{(0)} \psi_n^{(1)} >$$

$$= E_n^{(1)} \sum_{j=1}^{N} c_j < \phi_k \mid \phi_j > + E_n^{(0)} < \phi_k \mid \psi_n^{(1)} >$$

Just as in the nondegenerate development, the second term on the left hand side is the same as the second term on the right hand side, and they cancel. Because of orthogonality of the degenerate functions, the summation on the right hand side that is multiplied by the first-order correction to the energy simplifies to $c_k$.

$$\sum_{j=1}^{N} < \phi_k \mid \hat{H}^{(1)} \phi_j > c_j = c_k E_n^{(1)}$$

This expression may be rewritten by using the matrix repesentation of the perturbing Hamiltonian in the $\phi$ basis.

$$\sum_{j=1}^{N} \mathbf{H}_{kj}^{(1)\text{-}\phi} c_j = c_k E_n^{(1)}$$

And when all choices of k are considered, there are N equations of this form. They may all be collected into one matrix equation by arranging the c coefficients into a column vector (see Appendix I).

$$\mathbf{H}^{(1)\text{-}\phi} \vec{c} = \vec{c} E_n^{(1)} \tag{4-31}$$

This happens to be the standard form of a matrix eigenvalue equation.

The interpretation of Eq. (4-31) is that the first-order correction to the energy of the state of interest (n) is an eigenvalue of the matrix representation of the perturbing Hamiltonian in the basis of the degenerate functions $\{\phi_i\}$. The eigenvector in Eq. (4-31) is the set of coefficients that tells which linear combination of degenerate functions must represent the zero-order state function, that is, what values end up being used in Eq. (4-29). So, in the case of degenerate functions, a diagonalization procedure is used, just as in linear variation theory, except that the matrix representation is limited to just the functions that are degenerate with the state of interest. The diagonalization implies a transformation (see Appendix I) of the $\{\phi_i\}$ set of functions; they will be mixed with each other. If the eigenenergies obtained via Eq. (4-31) are all different, then the perturbation is said to have removed the degeneracy. After solving Eq. (4-31), the first-order correction to the wavefunction and all the higher order corrections may be obtained as in the nondegenerate case, if the degeneracy has been removed. Or if the degeneracy has not been removed, the expressions for higher order corrections are developed by allowing the corrections to the wavefunctions to be linear combinations of the degenerate functions, analogous to Eq. (4-29).

## 4.9 ANGULAR MOMENTUM OPERATORS AND EIGENFUNCTIONS

The importance of angular momentum in many quantum mechanical systems is its quantization. It is often the case that the angular momentum of a system will be restricted to certain values and not continuously variable. In the earliest days of quantum theories, the Bohr model of the hydrogen atom had some success in explaining atomic spectra by imposing a quantization condition on the electron's orbital angular momentum. This was a presumption, but the ultimate explanation of the mechanics of the hydrogen atom showed that indeed the electron's angular momentum is quantized. Today, many features of molecular spectra are understood in terms of angular momentum properties, and even reaction probabilities have been found that are dependent on the angular momenta of the colliding atoms or molecules.

The fundamental ideas of angular momentum in quantum systems draw on classical mechanical theory, and there one finds a notable correspondence between rectilinear and rotational motions. Rectilinear position and momentum are as basic as angular position and angular momentum. Rectilinear force has an analog in the torque about an axis, and in the appropriate equations of motion, the mass of a straight-moving body plays the same role as a rotating body's moment of inertia. In all respects, the analysis of the mechanics of rotating systems is not so much special as it is a generalization of the types of coordinates so as to include angular coordinates.

Angular momentum, like rectilinear momentum, is a vector quantity, and for a moving point-mass it may be defined as the vector cross-product of the position vector (from the axis of rotation to the mass) and the linear momentum vector.

$$\vec{L} = \vec{r} \otimes \vec{p} \qquad \Rightarrow \qquad \begin{aligned} L_x &= y\,p_z - z\,p_y \\ L_y &= z\,p_x - x\,p_z \\ L_z &= x\,p_y - y\,p_x \end{aligned} \qquad (4\text{-}32)$$

For a system of several point-mass particles, the total angular momentum is the vector sum of the angular momenta of each of the particles.

Quantum mechanical operators corresponding to the components of an angular momentum vector may be found directly from the familiar rectilinear position and momentum operators, e.g.,

$$\hat{L}_x = \hat{y}\,\hat{p}_z - \hat{z}\,\hat{p}_y$$

Therefore,

$$\begin{aligned} \hat{L}_x &= -i\hbar\left( y\frac{\partial}{\partial z} - z\frac{\partial}{\partial y} \right) \\ \hat{L}_y &= -i\hbar\left( z\frac{\partial}{\partial x} - x\frac{\partial}{\partial z} \right) \\ \hat{L}_z &= -i\hbar\left( x\frac{\partial}{\partial y} - y\frac{\partial}{\partial x} \right) \end{aligned} \qquad (4\text{-}33)$$

With these explicit forms for the operators, it is straightforward to show that the commutator of any pair of them produces the third one:

$$[\hat{L}_x, \hat{L}_y] = i\hbar\,\hat{L}_z$$

$$[\hat{L}_y, \hat{L}_z] = i\hbar \hat{L}_x \tag{4-34}$$

$$[\hat{L}_z, \hat{L}_x] = i\hbar \hat{L}_y$$

In practice, it is better to work with angular momentum in an angular coordinate system rather than a rectilinear system. This means using the spherical polar coordinate system defined as in Fig. 2.3 and transforming the operators in Eq. (4-33).

The coordinate transformation to and from spherical polar coordinates is

$$x = r \sin\theta \cos\phi \qquad r^2 = x^2 + y^2 + z^2$$

$$y = r \sin\theta \sin\phi \qquad \theta = \arccos(z/r)$$

$$z = r \cos\theta \qquad \phi = \arctan(y/x)$$

This transformation enables us to write the angular momentum component operators in either system. Chain rule differentiation provides the substitution for a differential operator in Eq. (4-33). For instance,

$$\frac{\partial}{\partial x} = \frac{\partial r}{\partial x}\frac{\partial}{\partial r} + \frac{\partial \theta}{\partial x}\frac{\partial}{\partial \theta} + \frac{\partial \phi}{\partial x}\frac{\partial}{\partial \phi}$$

and from the transformation equations,

$$\frac{\partial r}{\partial x} = \frac{x}{r} = \sin\theta \cos\phi, \qquad \text{and so on.}$$

If this is worked through completely, the following operator expressions are obtained.

$$\hat{L}_x = -i\hbar \left(-\sin\phi \frac{\partial}{\partial \theta} - \frac{\cos\phi}{\tan\theta}\frac{\partial}{\partial \phi}\right)$$

$$\hat{L}_y = -i\hbar \left(\cos\phi \frac{\partial}{\partial \theta} - \frac{\sin\phi}{\tan\theta}\frac{\partial}{\partial \phi}\right) \tag{4-35}$$

$$\hat{L}_z = -i\hbar \frac{\partial}{\partial \phi}$$

These can be applied to wavefunctions given in spherical polar coordinates to extract information about the angular momentum of a system.

Another useful operator may be found from the angular momentum component operators. The square of the angular momentum of a classical or a quantum mechanical system is a scalar quantity, since it is simply the dot product of $\vec{L}$ and itself. The dot product of any vector and itself equals the square of the length of the vector. Thus, $\vec{L} \bullet \vec{L}$ tells us the magnitude of the angular momentum because it corresponds to the square of that vector's length. The quantum mechanical operator that corresponds to this dot product is designated $\hat{L}^2$, and it can be applied to wavefunctions to find out about the magnitude of a system's angular momentum.

The explicit form of the $\hat{L}^2$ operator in any coordinate system may be derived from the component expression for a dot product; that is,

$$\hat{L}^2 = \hat{L}_x^2 + \hat{L}_y^2 + \hat{L}_z^2$$

In spherical polar coordinates, the following is obtained with Eq. (4-35).

$$\hat{L}^2 = -\hbar^2 \left( \frac{1}{\sin\theta} \frac{\partial}{\partial\theta} \sin\theta \frac{\partial}{\partial\theta} + \frac{1}{\sin^2\theta} \frac{\partial^2}{\partial\phi^2} \right) \tag{4-36}$$

This operator is related to the *Laplacian operator*, $\nabla^2$ (del squared), in spherical polar coordinates. From the definition of the Laplacian in Cartesian coordinates, and the coordinate transformation given above, we have that

$$\nabla^2 = \frac{\partial^2}{\partial x^2} + \frac{\partial^2}{\partial y^2} + \frac{\partial^2}{\partial z^2} \tag{4-37a}$$

$$= \frac{2}{r}\frac{\partial}{\partial r} + \frac{\partial^2}{\partial r^2} - \frac{\hat{L}^2/\hbar^2}{r^2} \tag{4-37b}$$

The Laplacian is used in the kinetic energy operator for a particle, because $\hat{p}^2/2m = -\hbar^2\nabla^2/2m$. Thus, the square of the angular momentum is intimately connected with the amount of kinetic energy of a moving particle.

*Spherical harmonic functions* are the functions of the two spherical polar coordinate angles, $\theta$ and $\phi$, that are eigenfunctions of the two operators $\hat{L}^2$ and $\hat{L}_z$. These functions are usually designated $Y(\theta,\phi)$, and the whole set is obtained by solving the two differential eigenequations,

$$\hat{L}^2 Y = \alpha Y$$

$$\hat{L}_z Y = \beta Y$$

There are an infinite number of solutions to these equations, and two integers are needed to label and distinguish the different spherical harmonic functions. The conventional choices for these two integers are $l$ and m. The $\theta$ dependence of these functions may be expressed with a special set of polynomials called the *associated Legendre polynomials.*

Associated Legendre polynomials for a variable z, designated $P_l^{|m|}(z)$, may be generated with the following two formulas.

$$P_l^0(z) = \frac{1}{2^l l!} \frac{d^l}{dz^l}\left[ (z^2 - 1)^l \right] \tag{4-38}$$

$$P_l^m(z) = (1 - z^2)^{m/2} \frac{d^m}{dz^m} P_l^0(z) \tag{4-39}$$

Notice that if $m > l$, these formulas lead to a function that is zero. Certain of these polynomials* are given explicitly in Table 4.1.

In the spherical harmonic functions, the "variable" for the Legendre polynomials is really a function, cos $\theta$. This means that after establishing the polynomials explicitly for z using Eqs. (4-38) and (4-39), the desired equations in terms of $\theta$ are obtained by substituting cos $\theta$ for z throughout. Trigonometric identities are used to simplify the expressions, and these forms are also presented in Table 4.1. The associated Legendre polynomials are orthogonal functions for the range $z = -1$ to $z = 1$. This corresponds to the range where $\cos\theta = -1$ to $\cos\theta = 1$, which is $\theta = 0$ to $\theta = \pi$. Over this range the functions are normalized by a simple factor that uses $l$ and m:

$$\sqrt{\frac{(2l + 1)\,(l - |m|)!}{2\,(l + |m|)!}}$$

The symbol $\Theta$ is used to designate a normalized associated Legendre polynomial in $\theta$; that is,

* The polynomials for $m = 0$, that is, those generated with only Eq. (4-38), are generally referred to as the Legendre polynomials, and then the superscript is suppressed. The polynomials that follow from Eq. (4-39) are the *associated* Legendre polynomials.

TABLE 4.1 Associated Legendre Polynomials Through $l = 4$.

| $P_l^0(z)$ | $P_l^{\lvert m \rvert}(z)$ | $P_l^{\lvert m \rvert}(\cos\theta)$ |
|---|---|---|
| $l = 0$ | $P_0^0 = 1$ | |
| $l = 1$ | $P_1^0 = z$ | $P_1^0 = \cos\theta$ |
| | $P_1^1 = \sqrt{1-z^2}$ | $P_1^1 = \sin\theta$ |
| $l = 2$ | $P_2^0 = (3z^2 - 1)/2$ | $P_2^0 = (3\cos^2\theta - 1)/2$ |
| | $P_2^1 = 3z\sqrt{1-z^2}$ | $P_2^1 = 3\sin\theta\cos\theta$ |
| | $P_2^2 = 3(1 - z^2)$ | $P_2^2 = 3\sin^2\theta$ |
| $l = 3$ | $P_3^0 = (5z^3 - 3z)/2$ | $P_3^0 = \cos\theta\,(5\cos^2\theta - 3)/2$ |
| | $P_3^1 = 3(5z^2 - 1)\sqrt{1-z^2}/2$ | $P_3^1 = 3\sin\theta\,(5\cos^2\theta - 1)/2$ |
| | $P_3^2 = 15z(1 - z^2)$ | $P_3^2 = 15\cos\theta\sin^2\theta$ |
| | $P_3^3 = 15(1 - z^2)^{3/2}$ | $P_3^3 = 15\sin^3\theta$ |
| $l = 4$ | $P_4^0 = (35z^4 - 30z^2 + 3)/8$ | $P_4^0 = (35\cos^4\theta - 30\cos^2\theta + 3)/8$ |
| | $P_4^1 = \sqrt{1-z^2}\,(35z^3 - 15z)/2$ | $P_4^1 = \sin\theta\cos\theta\,(35\cos^2\theta - 15)/2$ |
| | $P_4^2 = (1-z^2)(105z^2 - 15)/2$ | $P_4^2 = \sin^2\theta\,(105\cos^2\theta - 15)/2$ |
| | $P_4^3 = 105z(1-z^2)^{3/2}$ | $P_4^3 = 105\sin^3\theta\cos\theta$ |
| | $P_4^4 = 105(1-z^2)^2$ | $P_4^4 = 105\sin^4\theta$ |

$$\Theta_{lm}(\theta) = \sqrt{\frac{(2l+1)\,(l-|m|)!}{2\,(l+|m|)!}}\; P_l^{|m|}(\cos\theta) \tag{4-40}$$

Notice that the absolute value of m is used in Eq. (4-40), and so the $\Theta$ functions are the same for m and –m.

Spherical harmonics have a simple exponential dependence on the angle $\phi$ via functions designated $\Phi_m$.

$$\Phi_m(\phi) = e^{im\phi} / \sqrt{2\pi} \tag{4-41}$$

The spherical harmonics are products of $\Phi$ and $\Theta$ functions.

$$Y_{lm}(\theta,\phi) = \Theta_{lm}(\theta)\;\Phi_m(\phi)\;(-1)^{[m+|m|]/2} \tag{4-42}$$

The factor of $(-1)^{[m+|m|]/2}$ is an arbitrary phase factor that is introduced to conform to common conventions; its value is always 1 or –1. Notice that the $\Phi$ functions are not dependent on the $l$ integer. They are easily shown to be orthogonal and to be normalized for the range $\phi = 0$ to $\phi = 2\pi$. Thus, the spherical harmonic functions are a set of orthonormal functions over the usual ranges of angles in the spherical polar coordinate system. As a formula, this is

$$\int_0^{2\pi}\int_0^{\pi} Y^*_{lm}(\theta,\phi)\, Y_{l'm'}(\theta,\phi)\,\sin\theta\; d\theta\; d\phi = \delta_{ll'}\,\delta_{mm'} \tag{4-43}$$

This can be a very helpful expression for working with wavefunctions given in terms of $\theta$ and $\phi$.

With explicit forms of the spherical harmonic functions, it is possible to show that each is an eigenfunction of the operators $\hat{L}^2$ and $\hat{L}_z$ and that the eigenvalues come from $l$ and m.

$$\hat{L}^2\, Y_{lm} = l\,(l+1)\,\hbar^2\; Y_{lm} \tag{4-44}$$

$$\hat{L}_z\, Y_{lm} = m\hbar\, Y_{lm} \tag{4-45}$$

$l$ must be zero or a positive integer; solutions do not exist for other values. m is restricted because if its absolute value were greater than $l$, then the associated Legendre polynomial would be zero. This would give rise to a zero-valued spherical harmonic function that would not serve as a wavefunction since it would correspond to zero probability density everywhere. Therefore, $|m| \leq l$, or $m = -l, -l+1, \ldots, l-1, l$.

## 4.10 THE RIGID ROTATOR

An important model problem involving angular momentum is the rigid rotator. Two masses, $m_1$ and $m_2$, are taken to be connected by a massless rod that is absolutely rigid. Thus, the separation distance between the masses is fixed at some value, R, the length of the rod. Fig. 2.3 serves to depict this problem, provided we take the radial spherical polar coordinate, r, to have the fixed value R (i.e., r = R). After separating the motion corresponding to translation of the center of mass of the system, the quantum mechanical kinetic energy operator for the system in Fig. 2.3 is

$$\hat{T} = -\frac{\hbar^2}{2\mu}\nabla^2$$

where $\mu$ is the reduced mass and is given in the internal coordinates, r, $\theta$, $\phi$. Using Eq. (4-37b) but taking r to be fixed at the value R, we have

$$\hat{T} = \frac{\hat{L}^2}{2\mu R^2}$$

In this model situation, there is no potential of any sort, and so the kinetic energy operator and the Hamiltonian are one and the same.

The Schrödinger equation for the rigid rotator must be

$$\frac{\hat{L}^2}{2\mu R^2}\psi = E\psi$$

Since we know that the spherical harmonics are eigenfunctions of the $\hat{L}^2$ operator, then they are also eigenfunctions of the rigid rotator's Hamiltonian since it is proportional to $\hat{L}^2$. That is, the functions $Y_{lm}$ are the wavefunctions of the rigid rotator. The eigenenergies are obtained from Eq. (4-44):

$$\frac{\hat{L}^2}{2\mu R^2} Y_{lm} = l(l+1) \frac{\hbar^2}{2\mu R^2} Y_{lm}$$

$$\therefore \quad E_{lm} = l(l+1) \frac{\hbar^2}{2\mu R^2} \tag{4-46}$$

Notice that the energies increase quadratically with the quantum number $l$. Also, each energy level has a degeneracy of $(2l + 1)$, because there are $(2l + 1)$ choices of the m quantum number for any particular $l$; for all such choices, the eigenenergies are the same.

The physical picture of the rigid rotator is that of a dumbbell whirling about its center of mass. Its allowed states are those given by the allowed values of the two quantum numbers:

$l = 0, 1, 2, 3, \ldots$

$m = -l, \ldots, l$

The wavefunctions are eigenfunctions of the Hamiltonian, and also of $\hat{L}^2$ and of $\hat{L}_z$. The eigenvalues of $\hat{L}^2$ are as given in Eq. (4-44). On the basis of the postulates, we should expect that a measurement of the square of the angular momentum will produce one of these values. The square root of such a measurement is interpreted as the magnitude or length of the angular momentum vector. For a given $Y_{lm}$ state, this quantity is $\sqrt{l(l+1)}\,\hbar$. Thus, the only allowed lengths of the angular momentum vector are 0, $\sqrt{2}\,\hbar$, $\sqrt{6}\hbar$, $\sqrt{12}\hbar$, $\sqrt{20}\hbar$, and so on.

The eigenvalue of $\hat{L}_z$ for a $Y_{lm}$ function is $m\hbar$. This is the value of the z-component of the angular momentum vector. So, for each state of the rigid rotator, we know the length of the angular momentum vector and its z-component. However, we do not know the x- or y-component precisely since the wavefunctions are not eigenfunctions of the operators $\hat{L}_x$ and $\hat{L}_y$. The information we do have can be depicted by vectors indicating the possible orientations of the angular momentum vector with respect to the z-axis for the different possible lengths of the vector; this is shown in Fig. 4.4. So, our whirling quantum dumbell can "spin" only in fixed or discrete amounts and then only with the angular momentum vector at certain fixed angles.

The three operators, $\hat{L}^2$, $\hat{L}_z$, $\hat{H}$, commute in this model problem, because according to Theorem 4.4, if a set of functions exists that are

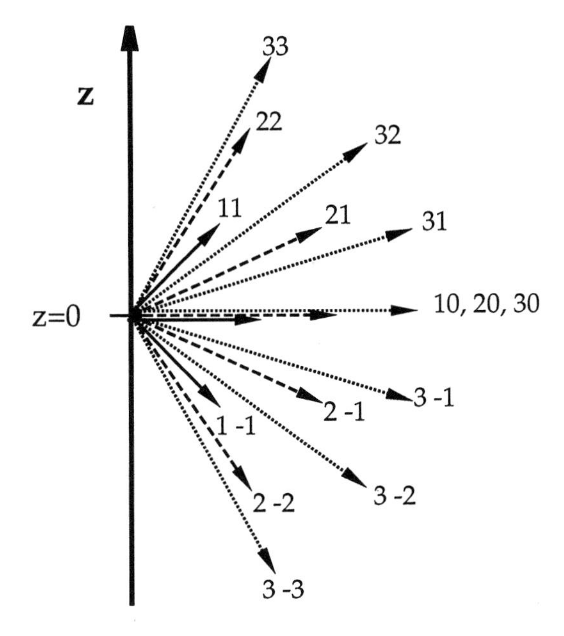


**FIGURE 4.4** The angular momentum vectors of the $Y_{lm}$ states of the rigid rotator for $l = 1$ (solid lines), $l = 2$ (dashed lines), and $l = 3$ (dotted lines). The arrows represent the orientation with respect to the z-axis for the particular $(l,m)$ state. The orientations with respect to the x-axis and the y-axis are not determined, and so this picture represents a planar slice through three-dimensional space such that the slice includes the z-axis and is perpendicular to the x-y plane. The angle between this planar slice and the x-axis or the y-axis, though, is arbitrary.

simultaneously eigenfunctions of any two operators, then those operators must commute. The spherical harmonic functions are eigenfunctions of each of these three operators.

The Hamiltonian must commute with $\hat{L}^2$ since it is proportional to $\hat{L}^2$. This may be seen by considering the commutator of some arbitrary operator $\hat{G}$ and that operator scaled by a constant g:

$$[\hat{G}, g\hat{G}] \equiv \hat{G}(g\hat{G}) - g\hat{G}\hat{G} = g(\hat{G}^2 - \hat{G}^2) = 0$$

The commutator of $\hat{L}^2$ with any of the angular momentum component operators happens to be zero, which means that $\hat{L}^2$ also commutes with

$\hat{L}_x$, $\hat{L}_y$ and $\hat{L}_z$. This can be shown using the explicit forms of these operators given in Eqs. (4-33) and (4-36), and it is a general result for any type of angular momentum. Another result that may also be obtained from using the explicit forms of the component operators is that they do not commute with each other. This is shown in Eq. (4-34). Therefore, the largest set of mutually commuting operators for the rigid rotator problem consists of three, the Hamiltonian, the $\hat{L}^2$ operator, and any one of the component operators.

It is significant that all of the component operators cannot be part of the mutually commuting set. If the component operators, the Hamiltonian, and $\hat{L}^2$ were all mutually commuting, then the corollary to Theorem 4.4 would say that a set of functions can be found that are simultaneously eigenfunctions of the Hamiltonian, of $\hat{L}^2$, and of all three components. In turn, that would say that we can measure each component of the angular momentum with no uncertainty, which would mean that for any state of the rigid rotator we can know exactly where the angular momentum vector points. That the component operators are *not* mutually commuting means that the wavefunctions of the system can be eigenfunctions of only one component operator, and that measurements with no uncertainty can be accomplished for only that one component; the other two will have nonzero uncertainty.

It is by convention that we choose the spherical harmonic functions such that they are eigenfunctions of the z-component operator and not the x- or y-component operator. But, there is nothing at all special about the z-direction in space. Rather, since we know that there will be one direction in which the angular momentum component is quantized (i.e., the wavefunctions will be eigenfunctions of *one* component operator), then the convention is that that direction, whatever it happens to be, becomes the z-direction.

## 4.11 COUPLING OF ANGULAR MOMENTA

In many chemical problems, there are multiple sources of angular momenta. For instance, electrons orbiting about a molecular axis may give rise to angular momentum at the same time that the rotation of the molecule as a whole about its mass center gives rise to angular momentum. Classically and quantum mechanically, angular momenta add vectorially, and the total angular momentum of a closed system is conserved. In the classical world, we may know the orientation and length of a given angular momentum vector at any instant in time. And so we may add together all such vectors to know the total angular momentum vector. In the quantum world, there is uncertainty in the orientation of an angular momentum vector, and so the addition or coupling of momenta calls for a different sort of analysis.

In a situation where some source of angular momentum is quantized, the associated angular momentum operator commutes with the Hamiltonian. The wavefunctions, then, are simultaneous eigenfunctions. Because of this, we may approach problems involving angular momentum, and we may develop general rules, without considering a specific problem or a specific Hamiltonian. In other words, much can be learned from considering the angular momentum features of quantum states on their own. An extension of the bra-ket notation convention is an aid in this endeavor. Instead of placing a function designation (e.g., $\psi$) in a bra or ket, we will place angular momentum quantum numbers for all angular momentum operators for which the wavefunctions must be eigenfunctions (e.g., $| l \; m >$ for a spherical harmonic function instead of $|Y_{lm}>$). Of course in a real problem, there may be a number of states with the same angular momentum quantum numbers but different energies, and so we will not be distinguishing among them in this analysis. Another common practice is to use J in the same way that L has already been used, except that J may represent *any* type of angular momentum, whereas L is usually reserved for orbital angular momentum.

Rules for adding angular momentum need only be developed for two sources at a time. The addition of three sources can be accomplished by adding two together and then adding that result to the third angular momentum vector. So, consider two independent sources, $\vec{J}_1$ and $\vec{J}_2$. In the absence of a physical interaction between the two sources, the wavefunctions are eigenfunctions of the total angular momentum operators for each source and of the z-component operators for each source. The designation of an angular momentum state in this situation requires four quantum numbers,

$$| j_1 \, m_1 \, j_2 \, m_2 >$$

$j_1$ and $j_2$ are the total angular momentum quantum numbers for the two independent sources, while $m_1$ and $m_2$ are the z-component quantum numbers. Thus, we may evaluate the effect of total source and z-component operators on the wavefunctions.

$$\hat{J}_1^2 \; | j_1 \, m_1 \, j_2 \, m_2 > \; = \; j_1 (j_1 + 1) \hbar^2 \; | j_1 \, m_1 \, j_2 \, m_2 >$$

$$\hat{J}_2^2 \; | j_1 \, m_1 \, j_2 \, m_2 > \; = \; j_2 (j_2 + 1) \hbar^2 \; | j_1 \, m_1 \, j_2 \, m_2 >$$

$$\hat{J}_{1z} \; | j_1 \, m_1 \, j_2 \, m_2 > \; = \; m_1 \hbar \; | j_1 \, m_1 \, j_2 \, m_2 >$$

$$\hat{J}_{2z} \; | j_1 \, m_1 \, j_2 \, m_2 > \; = \; m_2 \hbar \; | j_1 \, m_1 \, j_2 m_2 >$$

The possible values for the quantum number $m_1$ are $-j_1, \ldots, j_1$, and by counting those possibilities, we find there are $(2j_1 + 1)$ of them. This is the multiplicity of states from the $j_1$ angular momentum. The total number of states for the whole system is a product of multiplicities, $(2j_1 + 1)(2j_2 + 1)$.

Coupling of angular momenta comes about through an interaction that is brought into the Hamiltonian of the problem. An interaction that couples momenta must involve both angular momentum vectors. So, the form is generally that of a dot product of vectors, e.g. $\vec{J}_1 \bullet \vec{J}_2$. Such an interaction implies an energetic preference in how the two angular momentum vectors combine in forming a resultant vector. The possibilities, though, are restricted by the quantization of the total angular momentum and its z-component. For instance, the maximum possible z-component of the total angular momentum cannot be any more than the sum of the maximum possible z-components of the two vectors being combined. From Fig. 4.4, the maximum z-component for any angular momentum vector occurs when the associated m quantum number is equal to the j quantum number. Therefore, we may state a rule:

$$\text{Maximum resultant z-component} = [\text{maximum value of } m_1 + \text{maximum value of } m_2]\hbar$$

$$= (j_1 + j_2)\,\hbar$$

This must also equal the maximum z-component quantum number of the resultant vector, designated M, times $\hbar$. That is,

$$\text{Maximum value of M} = j_1 + j_2$$

Since any z-component quantum number can take on only certain values because of the length of the associated vector (i.e., $M = -J, \ldots, J$), then knowing the maximum value of M means knowing J. Thus, we conclude that one possible value for the resultant J quantum number is $j_1 + j_2$.

There are other possible resultant J quantum numbers. Notice that there are two ways in which a resultant vector may be composed for which $M = j_1 + j_2 - 1$. One way is with the first vector arranged so that $m_1 = j_1 - 1$, while for the second, $m_2 = j_2$. This is illustrated in Fig. 4.5. The second way is with $m_1 = j_1$, but $m_2 = j_2 - 1$. One $\hbar$ less in the resultant z-component comes about through addition of one or the other component vectors at an orientation with one $\hbar$ less in the z-component. These two arrangements of the source vectors do not distinctly correspond to resultant angular momentum states, but they do indicate that there will exist two states for which $M = j_1 + j_2 - 1$. One of those is expected, since it represents one of the possible choices of M with $J = j_1 + j_2$ (the possible J value already identified). The fact that there is another state with the same z-component means that there must be a state with a different J value, such that its maximum z-component is $(j_1 + j_2 - 1)\hbar$. That is, the next possible J value is $j_1 + j_2 - 1$.

If the process were continued to the next step down in the z-component of the resultant vector, it would turn out that there has to be another possible J value; this one being $j_1 + j_2 - 2$. And this pattern would continue up until $|j_1 - j_2| = 0$. Thus, the rule for adding angular momenta is that the possible values for the resultant or total J quantum number span a range given by the quantum numbers associated with the source vectors:

$$J = j_1 + j_2, j_1 + j_2 - 1, j_1 + j_2 - 2, \ldots, |j_1 - j_2| \tag{4-47}$$

The resultant states are eigenfunctions of the operators $\vec{J}^2$, $\vec{J}_1^{\,2}$, $\vec{J}_2^{\,2}$, and $\hat{J}_z$ and may be designated $|J\,M\,j_1\,j_2>$. They are not necessarily eigenfunctions of the z-component operators of the sources.

Incorporating a new interaction into a system's Hamiltonian, such as a coupling of angular momenta, should never change the number of states. As mentioned above, the number of states in the absence of any interaction

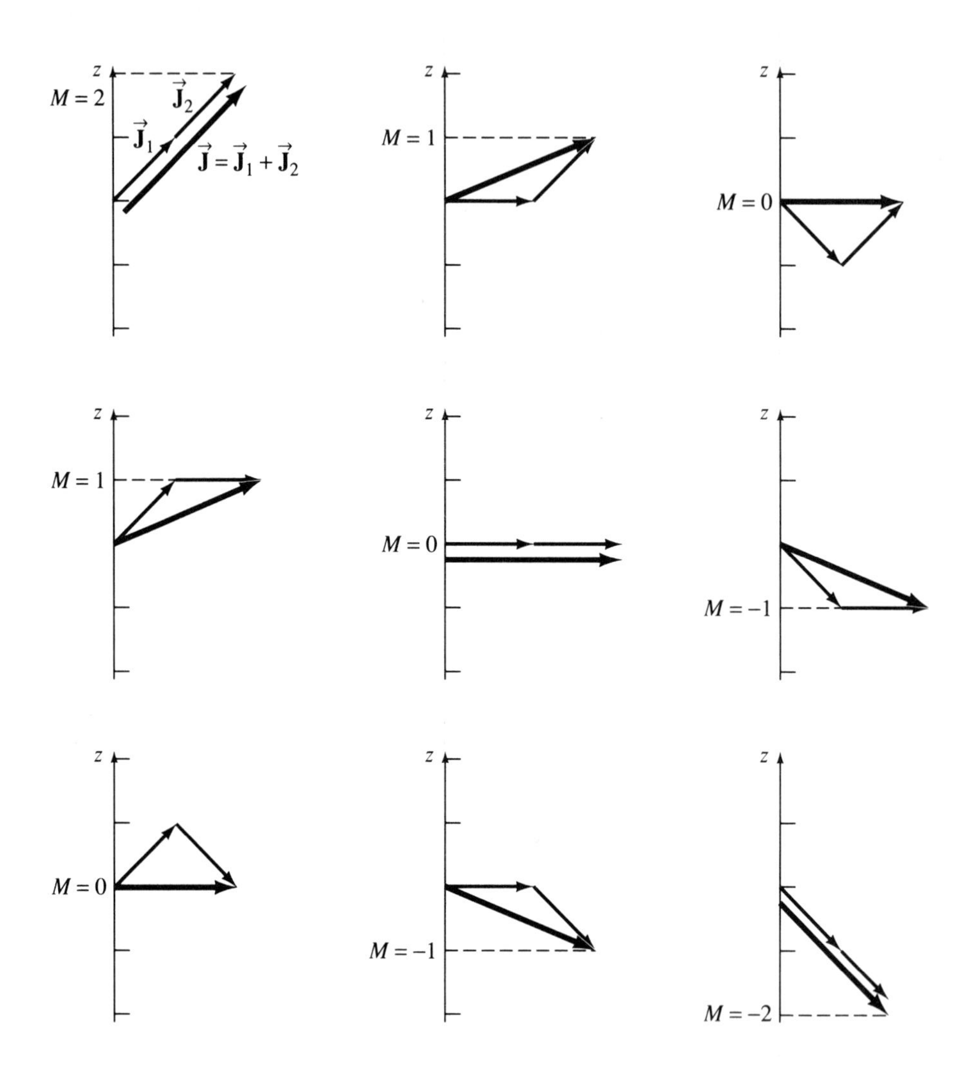


**FIGURE 4.5** For two angular momentum vectors, $\vec{J}_1$ and $\vec{J}_2$, with associated quantum numbers $j_1$ and $j_2$ both equal to 1, there will be $(2j_1 + 1)(2j_2 + 1) = 9$ different combinations of their possible orientations with respect to the z-axis. These nine ways are shown here, and for each a resultant vector from the sum of $\vec{J}_1$ and $\vec{J}_2$ has been drawn. In each case, the z-component of the resultant vector may be obtained from the sum of the z-component quantum numbers $m_1$ and $m_2$.

is a product of multiplicities, $(2j_1 + 1)$ $(2j_2 + 1)$. The number of states corresponding to the possible resultant J values in Eq. (4-47) is a sum of the multiplicities, $(2J + 1)$, for each possible value of J. It can be proven that this is the same number of states, i.e.,

$$(2j_1 + 1)(2j_2 + 1) = \sum_{J=|j_1-j_2|}^{j_1+j_2} (2J + 1)$$

This is demonstrated for a few examples. We specify values for the source momenta, $j_1$ and $j_2$, and find the resultant J's according to Eq. (4-47). The sum of the multiplicities for the possible J values is compared with the product of the $j_1$ and $j_2$ multiplicities.

| $j_1$ | $j_2$ | $(2j_1+1)(2j_2+1)$ | J | (2J+1) | Σ (2J+1) |
|---|---|---|---|---|---|
| 0 | 0 | **1** | 0 | 1 | **1** |
| 1 | 0 | **3** | 1 | 3 | **3** |
| 0 | 2 | **5** | 2 | 5 | **5** |
| 1 | 1 | **9** | 2 | 5 | |
| | | | 1 | 3 | |
| | | | 0 | 1 | **9** |
| 2 | 1 | **15** | 3 | 7 | |
| | | | 2 | 5 | |
| | | | 1 | 3 | **15** |
| 2 | 2 | **25** | 4 | 9 | |
| | | | 3 | 7 | |
| | | | 2 | 5 | |
| | | | 1 | 3 | |
| | | | 0 | 1 | **25** |

The J's in this list are the possible values according to Eq. (4-47).

An interesting and necessary feature of the coupled angular momentum states is that they are eigenfunctions of the coupling interaction operator, $\vec{J}_1 \bullet \vec{J}_2$. This can be shown by starting from an expression for the dot product of $\vec{J} = \vec{J}_1 + \vec{J}_2$ and itself.

$$\vec{J} \bullet \vec{J} = (\vec{J}_1 + \vec{J}_2) \bullet (\vec{J}_1 + \vec{J}_2)$$

$$= \vec{J}_1 \bullet \vec{J}_1 + 2\vec{J}_1 \bullet \vec{J}_2 + \vec{J}_2 \bullet \vec{J}_2$$

Rearranging terms yields

$$\vec{J}_1 \bullet \vec{J}_2 = [\vec{J} \bullet \vec{J} - \vec{J}_1 \bullet \vec{J}_1 - \vec{J}_2 \bullet \vec{J}_2]/2$$

$$= [\hat{J}^2 - \hat{J}_1^2 - \hat{J}_2^2]/2 \qquad (4\text{-}48)$$

This says that the operator corresponding to the dot product of the two source vectors is the same as a combination of the three operators that give the lengths of the angular momenta. The coupled states, $|J\ M\ j_1\ j_2>$, are eigenfunctions of these three operators, and so they must be eigenfunctions of the combination in Eq. (4-48).

$$(\vec{J}_1 \bullet \vec{J}_2)\,|J M j_1\, j_2> = [\hat{J}^2 - \hat{J}_1^2 - \hat{J}_2^2]/2\ |J M j_1\, j_2>$$

$$= [J(J+1) - j_1(j_1+1) - j_2(j_2+1)]\hbar^2/2\ |J M j_1\, j_2> \qquad (4\text{-}49)$$

Eq. (4-49) will be used later to evaluate the effect of Hamiltonian terms that take the form of a dot product of two angular momentum vectors. The states that are appropriate in the absence of a coupling interaction, $|\,j_1\ m_1\ j_2\ m_2>$, are not eigenfunctions of $\vec{J}_1 \bullet \vec{J}_2$.

## Exercises

1. Show that the n = 0, 1, and 2 wavefunctions of the harmonic oscillator are orthogonal to each other.
2. Determine whether or not the wavefunctions of the harmonic oscillator could also be eigenfunctions of the operators $\hat{x}$ and $\hat{p}_x$.

3. For the harmonic oscillator, which of the following are Hermitian operators, $\hat{x}$, $\hat{x}^2$, $\hat{p}$, and $\hat{p}^2$?

4. Find the number of degenerate states for the lowest five energy levels of a two-dimensional oscillator of the type shown in Fig. 4.1 given that $k_x = 4\,k_y$.

5. Find the degeneracy of the lowest four energy levels of an isotropic three-dimensional harmonic oscillator ($k_x = k_y = k_z$).

6. Given that for a certain problem it has been determined that the energy levels depend on two quantum numbers, n and m, according to the expression

$$E_{n,m} = -\frac{\hbar a}{n^2} + \frac{\hbar a m^2}{4}$$

where a is a constant. If n can have only the value 1, 2, or 3, and if m can be zero or any integer from $-(n+1)$ to $n+1$, find the degeneracies of all the energy levels.

7. Apply the variational method to the harmonic oscillator problem using the following as the trial wavefunction.

$$\Gamma(x) = N\,x\,e^{-\alpha x^2}$$

$\alpha$ is the adjustable parameter, and N is the normalization constant. N has a dependence on the value of $\alpha$ that must be found in order to work out this problem. How much higher is this energy than that of the true ground state? Explain this result.

8. Find the first-order perturbation theory correction to the energies of the ground and first two excited states of the harmonic oscillator in terms of the constant g in the perturbing Hamiltonian,

$$\hat{H}^{(1)} = gx^4$$

9. Find the first- and second-order perturbation theory corrections to the energy of the ground state of the harmonic oscillator in terms of the constant g in the perturbing Hamiltonian

$$\hat{H}^{(1)} = g\,x^3$$

10. Derive a general expression for the third-order perturbation theory correction to the energy of a quantum state.

11. Write the expression for the probability density function for a harmonic oscillator that has been prepared in the time-varying superposition of the ground and first excited time-independent states. That is,

$$\Psi(x,t) = \frac{1}{\sqrt{2}} \psi_{n=0}(x)\, e^{-i\,\omega t/2} + \frac{1}{\sqrt{2}} \psi_{n=1}(x)\, e^{-3\,i\,\omega t/2}$$

From this, determine the behavior of the probability density as a function of time at the specific point, x = 0.

12. Verify the derivation of the expressions in Eq. (4-34).

13. Derive Eq. (4-37b) from Eq. (4-37a).

14. Show that $\hat{L}^2$ commutes with $\hat{L}_x$ and $\hat{L}_y$ by using the explicit spherical polar coordinate forms.

15. Show that for any integer m, the function $\sin^m\theta \cos\theta\, e^{i\,m\phi}$ is an eigenfunction of $\hat{L}^2$. Find the eigenvalue.

16. Find the commutator of $\hat{L}_x$ and $\hat{L}_y$ in spherical polar coordinates.

17. Write out the explicit form of $Y_{54}(\theta,\phi)$.

18. Evaluate $< Y_{20} \mid \hat{L}_x Y_{20} >$ and $< Y_{20} \mid \hat{L}_x^2 Y_{20} >$.

19. Verify Eq. (4-43) for $l = 2$, m = 1 and $l' = 1$, m' = 1.

20. What are the values of the total angular momentum quantum number J for a problem with three coupled sources of angular momentum, $j_1 = 2$, $j_2 = 1$, and $j_3 = 1$?

21. If it were possible to have an angular momentum source with an angular momentum quantum number of 1/2, what would be the values of the total angular momentum quantum number J for a system of two such sources coupled together? And of three, and four?

22. Apply the variational method to the harmonic oscillator problem using $c_1$ and $c_2$ as the adjustable parameters in the trial wavefunction

$$\phi = \left\{ c_1\, \beta x + c_2\, (\beta^2 x^2 - 1/2) \right\} e^{-\beta^2 x^2/2}$$

Compare with the exact wavefunctions and account for the result.

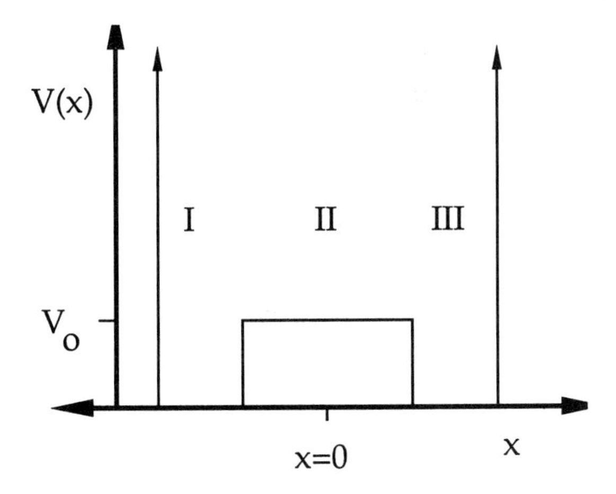


**FIGURE 4.6** A particle-in-a-box problem, but with a barrier step potential. The potential is taken to be infinite, except in regions I, II, and III. In regions I and III, the potential is zero. In region II, the potential is the constant value $V_0$.

23. The particle-in-a-sloped-box problem is the usual particle-in-a-box problem but with the bottom of the box sloped because of a linear potential of the form $V(x) = ax$ inside the box. Treat this as a perturbation on the usual particle-in-a-box problem and find the energies and wavefunctions correct to first order. Also, find the second-order energy corrections.

24. If the lowest energy state of the particle-in-a-box problem shown in Fig. 4.6 were at an energy slightly greater than $V_0$, in what ways would the wavefunctions of the lowest and first excited states differ from the wavefunctions of the same system but in the absence of the barrier step potential? Use a sketch to show qualitative features.

25. Find the first-order energy corrections for the first four states of the particle-in-a-box problem if there is a perturbing potential of the form

    $V'(x) = 0.1 \sqrt{\frac{2}{l}} \sin(\pi x / l)$.

26. For a particle in a three-dimensional box with equal length sides in the x-, y-, and z-directions, find the degeneracy for each of the first five energy levels.

27. Find the degeneracy of the first four energy levels of a particle in a three-dimensional box with the lengths of the sides related as $l_x = 16l_y = 4l_z$.

## Bibliography


1. E. Kreyszig, *Advanced Engineering Mathematics* (John Wiley and Sons, New York, 1967).
2. H. G. Hecht, *Mathematics in Chemistry: An Introduction to Modern Methods* (Prentice-Hall, Englewood Cliffs, New Jersey, 1990).
3. H. Margeneau and G. M. Murphy, *The Mathematics of Physics and Chemistry* (Van Nostrand and Co., Princeton, New Jersey, 1956).

(Also, see list of introductory texts in Chap. 3)

*INTERMEDIATE AND ADVANCED LEVEL TEXTS:*

4. P. W. Atkins, *Molecular Quantum Mechanics*, 2nd ed. (Oxford University Press, New York, 1983).
5. I. Levine, *Quantum Chemistry* (Allyn and Bacon, Boston, 1970).
6. E. Merzbacher, *Quantum Mechanics*, 2nd ed. (John Wiley and Sons, New York, 1970).
7. C. S. Johnson, Jr., and L. G. Pedersen, *Problems and Solutions in Quantum Chemistry and Physics* (Addison-Wesley, Reading, Massachusetts, 1974).
8. M. E. Rose, *Elementary Theory of Angular Momentum* (John Wiley and Sons, New York, 1957).
9. R. N. Zare, *Angular Momentum* (John Wiley and Sons, New York, 1988).

# Chapter 5

# Vibrational-Rotational Spectroscopy

*Quantum mechanical fundamentals are applied to the vibrational and rotational motions of molecules in this chapter. The analysis begins with an idealization of a typical diatomic, and then continues to a more realistic treatment. The resulting picture provides the basis for interpreting infrared and microwave spectra, and this chapter explores the way in which structural and energetic information is deduced from these spectra.*

## 5.1 MOLECULAR SPECTROSCOPY

Molecular spectroscopy refers to a wide category of experiments that involve probing individual molecules often by means of electromagnetic radiation. Radiation is used to induce transitions between molecular eigenstates. The process of absorbing or emitting radiation must conserve energy, and so the quantum of energy available in a photon,

$$E = h\nu$$

must usually match the energy change in the molecule. That is, for some frequency of radiation $\nu$, the quantity $h\nu$ will equal the *difference* in energy between two particular eigenstates of a system. A sample may absorb radiation at specific frequencies that correspond to eigenenergy differences, and so if we simultaneously vary the frequency of applied radiation and monitor the intensity of the radiation after passing through a sample, we will identify *transition frequencies* as drops in intensities. Thus, monitoring the radiation in a spectroscopic experiment, directly or indirectly, is a means of investigating the energies of the states of molecules.

The transition frequencies of a molecule in some region of the electromagnetic spectrum (e.g., visible, infrared) give information on the values of eigenenergy differences. Quantum theory relates these values to molecular features such as the stiffness of different chemical bonds and the molecular structure. Furthermore, the intensity, apart from serving as a qualitative indicator (absorbance at $\nu$ vs. non-absorbance), may be related to the quantum features of the molecule. But this will be discussed only after considering how transition frequencies relate to energy levels and molecular features.

Spectroscopic investigation of molecules was a revolution in chemistry that took place in the middle of this century. It has provided the detailed picture of molecular structure, molecular dynamics, and molecular properties upon which numerous principles of synthesis, molecular biology, and materials science have been established. New spectroscopic experiments are continuing to be devised, aided partly by technological advances in electronics, and now very subtle molecular features are being characterized.

A basic spectroscopic experiment involves a light source that can be focused on and passed through a sample. Detectors are required to monitor the light intensity before and after the sample. If the sample absorbs some of the impinging radiation, then the light intensity drops upon passing through the sample. The basic experiment yields a relation between the light intensity and the frequency of the light through some range of frequencies. Unfortunately, there is nothing that directly tells which states are associated with a given energy difference. Thus, the task of analyzing a spectrum usually starts with assigning frequencies of absorption to specific pairs of states. The job of making these assignments also involves the use of quantum mechanics so as to find out what types of states and energies are possible.

## 5.2 VIBRATION AND ROTATION OF A DIATOMIC

A widely used type of spectroscopy, infrared spectroscopy, probes the vibrations and rotations of simple molecules. The first such case to consider is that of a diatomic because the mechanics are just those of a two-body system. After separating out translational motion, the Hamiltonian is the kinetic energy operator, conveniently expressed in spherical polar coordinates (Fig. 2.3), plus a potential that depends only on the separation of the atoms. The potential, V(r), is something that develops because of the electrons and the chemical bonding in which they participate. A later chapter will examine the quantum mechanics behind bonding and the origin of the internuclear potentials, V(r), for diatomics. For now, we approach the quantum mechanics of a diatomic molecule by leaving the potential unspecified.

The Schrödinger equation for a vibrating and rotating diatomic, or generally for a two-body problem, with $\mu$ being the reduced mass, is

$$-\frac{\hbar^2}{2\mu}\nabla^2\,\psi(r,\theta,\phi) \;+\; V(r)\,\psi(r,\theta,\phi) \;=\; E\,\psi(r,\theta,\phi) \qquad (5\text{-}1)$$

With Eq. (4-37b), the operator $\nabla^2$ may be expressed in spherical polar coordinates and in terms of the angular momentum operator, $\hat{L}^2$.

$$-\frac{\hbar^2}{2\mu}\left(\frac{2}{r}\frac{\partial\psi}{\partial r} + \frac{\partial^2\psi}{\partial r^2}\right) + \frac{\hat{L}^2}{2\mu r^2}\,\psi \;+\; V(r)\,\psi \;=\; E\,\psi \qquad (5\text{-}2)$$

This differential equation is separable into radial and angular parts, where the angular wavefunctions must be the spherical harmonics. That is, if we assume a product form for the eigenfunction, $\psi$,

$$\psi(r,\theta,\phi) \;=\; Y_{lm}(\theta,\phi)\,R(r) \qquad (5\text{-}3)$$

then applying the operator $\hat{L}^2$ (and the factor $2\mu r^2$) produces

$$\frac{\hat{L}^2}{2\mu r^2}\,\psi \;=\; \frac{l(l+1)\hbar^2}{2\mu r^2}\,Y_{lm}\,R \qquad (5\text{-}4)$$

Therefore, with the product form of $\psi$ of Eq. (5-3) substituted into the Schrödinger equation, Eq. (5-2), we have

$$-\frac{\hbar^2 Y_{lm}}{2\mu}\left(\frac{2}{r}\frac{\partial R}{\partial r} + \frac{\partial^2 R}{\partial r^2}\right) + \frac{l(l+1)\hbar^2}{2\mu r^2} Y_{lm} R + V(r) Y_{lm} R = E Y_{lm} R$$

Dividing this equation by the spherical harmonic function, $Y_{lm}$, establishes the separability of the Schrödinger equation because the resulting differential equation is only in terms of the variable r.

$$-\frac{\hbar^2}{2\mu}\left(\frac{2}{r}\frac{\partial R}{\partial r} + \frac{\partial^2 R}{\partial r^2} - \frac{l(l+1)}{r^2} R\right) + V(r) R = E R \qquad (5\text{-}5)$$

This is called the radial Schrödinger equation of the two-body problem.

Though the two-body Schrödinger equation is separable, that is, $\psi(r,\theta,\phi) = R(r)Y_{lm}(\theta,\phi)$, it is important that the radial equation, Eq. (5-5), is still connected with the angular part of the problem. It incorporates the angular momentum quantum number $l$, and so Eq. (5-5) really represents an infinite number of differential equations corresponding to the infinite possible choices of $l$ (namely, 0, 1, 2, . . . ). Thus, any radial function, R(r), found from solving Eq. (5-5) will have to be labelled with the $l$ value.

At this point, it becomes convenient to rewrite Eq. (5-5) in terms of a new variable, s, which is the displacement from some fixed radial separation.

$$s \equiv r - r_e \qquad (5\text{-}6)$$

$r_e$ is a fixed value, and it might be chosen to be the equilibrium value of r or the separation of the two bodies at the potential minimum. Then, s = 0 corresponds to $r = r_e$, the potential minimum. On a graph of V versus s, the point s = 0 is the bottom of the potential well. It is also convenient to introduce a new function, $S$, defined in terms of the still undetermined function R.

$$S(s) = S(r - r_e) \equiv r R(r) \qquad (5\text{-}7)$$

Since the differential in s equals the differential in r [i.e., ds = dr from Eq. (5-6)], then differentiating $S(s)$ with respect to s must be the same as differentiating rR(r) with respect to r. Doing this twice produces the following relation between $S$ and R.

$$\frac{\partial^2 S}{\partial s^2} = r\left(\frac{2}{r}\frac{\partial R}{\partial r} + \frac{\partial^2 R}{\partial r^2}\right) \tag{5-8}$$

The right hand side is a part of Eq. (5-5), and so with the appropriate substitutions, and then with multiplication by r, Eq. (5-5) is now written in the following way.

$$-\frac{\hbar^2}{2\mu}\left[\frac{\partial^2 S(s)}{\partial s^2} - \frac{l(l+1)}{(s+r_e)^2} S(s)\right] + V(s+r_e)\, S(s) = E\, S(s) \tag{5-9}$$

This is the radial Schrödinger equation expressed in terms of $S(s)$ instead of R(r).

We may immediately consider the special case of Eq. (5-9) when the angular momentum is zero, that is, $l = 0$. The Schrödinger equation is then

$$-\frac{\hbar^2}{2\mu}\frac{\partial^2 S(s)}{\partial s^2} + V(s+r_e)\, S(s) = E\, S(s) \tag{5-10}$$

This equation is to be solved, but only upon specification of the potential function, V. One possibility is for the potential to be harmonic, $V(s+r_e) = ks^2/2$. Notice that then Eq. (5-10) becomes identical with the standard harmonic oscillator problem of Eq. (3-8), except that here the variable has been named s. A harmonic potential is the simplest, somewhat realistic choice for the stretching potential of a chemical bond. However, the harmonic form must be recognized as an approximation or as an idealization of the true molecular potential because, unlike a harmonic function, the potential for lengthening a chemical bond must reach a plateau or a constant potential energy once the chemical bond has completely broken and the atoms are not interacting. The harmonic function $kx^2/2$, on the other hand, goes to infinity as x gets larger; this "harmonic chemical bond" never breaks. Furthermore, even in the equilibrium region, there is no fundamental reason that the potential function should have precisely the shape of a parabola. But in many cases, a harmonic approximation is a fairly good approximation for the lower energy states. To be more realistic about the potential requires a more general functional form, such as a higher order polynomial. This is discussed further on.

The cases when $l > 0$ in Eq. (5-9) present only a small complication of the basic treatment because the possibly troublesome $l(l+1)$ term can be combined with the potential. That is, we may regard the potential as a combination of the $l(l+1)$ term with the original potential.

$$V_l^{eff}(s + r_e) \equiv V(s + r_e) + \frac{l(l+1)\hbar^2}{2\mu (s + r_e)^2} \tag{5-11}$$

Of course, instead of one potential, there is now a potential for each choice of $l$. Any one of these new potentials is termed an *effective potential* because the effect of the angular momentum term has been built into it.

Next, consider a power series expansion of the s dependence of the effective potential in Eq. (5-11), i.e., $(s + r_e)^{-2}$.

$$\frac{1}{(s + r_e)^2} = \frac{1}{r_e^2} - \frac{2s}{r_e^3} + \frac{6s^2}{r_e^4} - \frac{24 s^3}{r_e^5} + \dots \tag{5-12}$$

This, of course, is an infinite-order polynomial in s, and if the fundamental potential, V, is a polynomial as well, then the pieces on the right hand side of Eq. (5-11) can be combined term by term. An approximation of the expansion in Eq. (5-12) may be made by truncating it. The validity of the approximation depends on how large s may get relative to $r_e$. If s is likely to be less than 10% of $r_e$, for instance, then the third term will be only three-tenths of the second term, the fourth term will be only four-tenths of the third term, and so on. The terms will quickly become tiny, and truncation after a few should serve as a high-quality approximation. For typical chemical bonds in diatomic molecules, the classical turning point, which characterizes the extent of displacement from equilibrium, is often on the order of or less than 10% of the equilibrium bond length: A truncated power series is a good approximation, normally.

The most drastic truncation of the expansion in Eq. (5-12) is to keep only the first term. Using this approximation in the diatomic's vibration-rotation Hamiltonian corresponds to a neglect of the coupling of the vibrational motion with the rotational motion. It would be as if the system were effectively a rigid rotator, even though its ongoing vibration makes it anything but rigid. With this drastic approximation, Eq. (5-9) becomes

$$-\frac{\hbar^2}{2\mu}\left[\frac{\partial^2 S(s)}{\partial s^2} - \frac{l(l+1)}{r_e^2} S(s)\right] + V(s + r_e)\, S(s) = E\, S(s)$$

It is not necessary to rewrite this with $V^{eff}$ since the constant operator term may be brought to the right hand side and combined with E.

$$-\frac{\hbar^2}{2\mu}\frac{\partial^2 S(s)}{\partial s^2} + V(s + r_e)\, S(s) = E'\, S(s)$$

$$\text{where} \quad E' \equiv E - \frac{\hbar^2 l(l+1)}{2\mu r_e^2}$$

If V were harmonic, this would again be the Schrödinger equation of the harmonic oscillator, and E' would be found to be the harmonic oscillator energies.

From knowing the solutions of the harmonic oscillator problem, we arrive at the simplest idea of the energies, E, of the rotating-vibrating diatomic system. Using J instead of $l$ for the angular momentum quantum number, because that is the convention for molecular rotation, the possible state energies for the diatomic whose potential has been approximated as harmonic and for which vibration-rotation interaction is neglected are given as

$$E_{nJ} = (n + \frac{1}{2})\,\hbar\,\omega + \frac{\hbar^2 J(J+1)}{2\mu r_e^2} \qquad (5\text{-}13)$$

The energy levels are given by the vibrational quantum number, n, and the rotational quantum number, J. The approximations made to arrive at this result are checked, in the end, by comparing energies predicted from Eq. (5-13) with those measured in a spectroscopic experiment, and we shall consider certain specific data later.

The typical sizes of the constants in Eq. (5-13) are such that a change in the vibrational quantum number, n, will affect the energy more than a change in the rotational quantum number. For instance, for the molecule hydrogen fluoride, the quantity $\hbar\omega$ is around 4140 $cm^{-1}$, whereas $\hbar^2/(2\mu r_e^2)$ is only 21 $cm^{-1}$. Thus, the energy will be 4140 $cm^{-1}$ greater if n is increased from zero to one, but only 42 $cm^{-1}$ greater if J is increased from zero to one. A schematic representation of the associated energy level pattern is given in Fig. 5.1.

The truncation of Eq. (5-12) can be made to include more terms. With the truncation after two terms, the improved approximation is

$$\frac{1}{(s + r_e)^2} \cong \frac{1}{r_e^2} - \frac{2s}{r_e^3}$$

The right hand side is a first-order polynomial in s. So, within this approximation, the effective potential of Eq. (5-11) has the form

$$V^{eff}(s + r_e) = V(s + r_e) + a + b\,s$$

And if the original potential is harmonic, or taken to be harmonic, then the effective potential is

$$V^{eff}(s + r_e) = a + b\,s + \frac{1}{2}k\,s^2 \tag{5-14}$$

$$\text{where} \quad a = \frac{\hbar^2 J(J+1)}{2\mu r_e^2} \quad \text{and} \quad b = -\frac{2a}{r_e}$$

Of course, this is still a harmonic potential, since it is a quadratic polynomial in s. As shown in Fig. 5.2, it is a parabola differing from the parabola of $V(s + r_e)$ in the position of its minimum. Instead of the minimum being $V(r_e)=0$, the minimum is at some position we will call $\delta$, and then $V^{eff}(\delta + r_e) = V_o$, where the constant $V_o$ is not necessarily zero.

The effect of shifting the minimum of the parabolic potential of a harmonic oscillator changes the energies of the eigenstates in only one way. The energy amount that the parabola vertically shifts simply adds to each eigenenergy. For instance, wavefunctions that satisfy

$$-\frac{\hbar^2}{2\mu}\frac{d^2\psi_n}{dx^2} + \frac{1}{2}k\,x^2\,\psi_n = (n + \frac{1}{2})\,\hbar\omega\,\psi_n$$

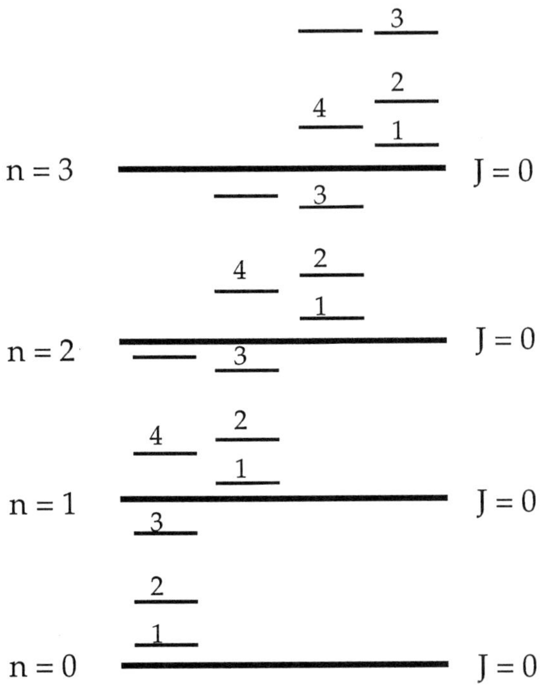

n = 3
n = 2
n = 1
n = 0
J = 0
J = 0
J = 0
J = 0
1 2 3 4
1 2 3 4
1 2 3 4
1 2 3

**FIGURE 5.1** The energy levels of a vibrating-rotating diatomic molecule according to Eq. (5-13). Each horizontal line in this sketch represents an energy level with energy increasing in the vertical direction. The levels are labelled by the vibrational quantum number, n, and the rotational quantum number, J. The long, thick lines are all J = 0 levels, and the energy spacing between any pair of these lines is $\hbar\omega$. The diagram has been terminated at n = 3, but the energy levels continue on infinitely in the same pattern. Levels for which J is not zero have been drawn as short, thin lines to help organize the drawing, but there is nothing fundamentally different between these and the J = 0 levels. These lines appear as stacks or manifolds originating from each J = 0 line. The leftmost manifold is for all the states for which n = 0, and the rightmost manifold is for all those for which n = 3. The numbers above four of these lines show the sequence of the J quantum numbers. Higher J levels exist, but have not been drawn. The degeneracy of each level is (2J + 1).

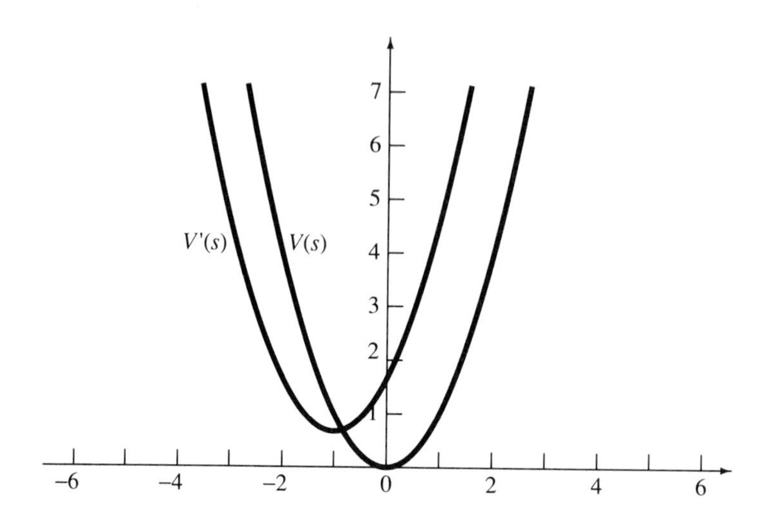


**FIGURE 5.2** A plot of the function $V(s) = s^2/2$ and of the function $V'(s) = 1 + s + s^2/2$. Both functions are parabolas with the same curvature; however, the minimum of V'(s) has been shifted by an amount $\delta = -1$, and the value of the potential at that point is $V'(\delta) = 1/2$ rather than zero.

---

have to satisfy the Schrödinger equation,

$$-\frac{\hbar^2}{2\mu}\frac{d^2\psi_n}{dx^2} + (V_o + \frac{1}{2}kx^2)\,\psi_n = \left[ V_o + (n + \frac{1}{2})\,\hbar\omega \right]\psi_n$$

with $V_o$ being a constant. The eigenenergies here differ only by the constant $V_o$. So, for the harmonic oscillator with the effective potential of Eq. (5-14), all the energy levels are shifted by whatever energy the minimum of the potential has been shifted. (If the x-value of the minimum is shifted, there is no change in the eigenenergies because the force constant is the same; the curvature of the parabola is unaffected by displacement of the minimum to the left or right.) Therefore, the effect of the better approximation of Eq. (5-12) that we are considering, can be determined by finding the location of the minimum of the effective potential in Eq. (5-14).

That means finding $\delta$ and $V_o$, as defined above. $\delta$ is the value of s when the first derivative of $V^{eff}$ is zero, and so

$$\delta = -\frac{\hbar^2 J(J+1)}{k \mu r_e^3} \tag{5-15}$$

$V_o$ is the value of $V^{eff}$ when s = $\delta$, and so

$$V_o = -\frac{\hbar^4 [J(J+1)]^2}{2 k \mu^2 r_e^6} + \frac{\hbar^2 J(J+1)}{2 \mu r_e^2} \tag{5-16}$$

Thus, the energy levels are now those of the harmonic oscillator but with the constant, $V_o$, added:

$$E_{nJ} = (n + \frac{1}{2}) \hbar\omega + \frac{\hbar^2 J(J+1)}{2 \mu r_e^2} - \frac{\hbar^4 [J(J+1)]^2}{2 k \mu^2 r_e^6} \tag{5-17}$$

The difference between this and Eq. (5-13) is just the last term. This last term is said to be associated with the effect of centrifugal distortion.

The physical idea of centrifugal distortion is that rotational motion contributes to the stretching of the "spring" between the particles because of centrifugal force. Thus, the separation distance, which is taken to be fixed at $r_e$ in the most drastic approximation of Eq. (5-12), will actually become somewhat longer as J increases. This means that the actual rotational energy will be less than what Eq. (5-13) dictates. The new term in Eq. (5-17) provides the kind of downward correction to the energies that is expected to arise from the centrifugal distortion of the bond. Notice that the J dependence of this term is quartic; it will diminish the energies of high-J states much more than low-J states.

Further improvements can be made in the vibrational-rotational energy level expression for a diatomic by carrying through the next (third) term in the truncation of Eq. (5-12). This term is quadratic in the displacment coordinate, s, and so its inclusion implies a different force constant in the effective potential.

$$V_J^{eff}(s + r_e) = V(s + r_e) + \frac{J(J+1)\hbar^2}{2\mu} \left( \frac{1}{r_e^2} - \frac{2s}{r_e^3} + \frac{6s^2}{r_e^4} \right) \tag{5-18}$$

The J(J+1) term of this potential is treated as a perturbation of the non-rotating harmonic oscillator. The part of the perturbation that goes as $1/r_e^2$ is exactly treated at first order because it is a constant term. It turns out that the second-order energy correction from the part of the perturbation that goes as $-2s/r_e^3$ is the same as the exact result. So, with second order as a chosen point of truncation of the perturbative corrections to the energy, the complete energy level expression [using the entire J(J+1) term in Eq. (5-18) as the perturbation] is,

$$E_{n,J} = \hbar\omega\left(n + \frac{1}{2}\right) + \frac{\hbar^2 J(J+1)}{2\mu r_e^2}$$

$$- \frac{\hbar^4 J^2 (J+1)^2}{2k\mu^2 r_e^6} + \frac{3\hbar^2}{\omega\mu^2 r_e^4}\left(n + \frac{1}{2}\right) J(J+1) \tag{5-19}$$

The third term is the same as that in Eq. (5-17) and is associated with centrifugal distortion. The last term involves both the vibrational quantum number, n, and the rotational quantum number, J, and may be regarded as a manifestation of the coupling of rotational motion with vibrational motion.

## 5.3 VIBRATIONAL ANHARMONICITY

*Vibrational anharmonicity* refers to the effects on energy levels of an otherwise harmonic oscillator from those parts of the stretching potential that are anharmonic, that is, those that do not vary as the square of the displacement coordinate. The most realistic vibrational potentials of molecules are not strictly harmonic, and so for a most precise quantum mechanical treatment there must be an incorporation of anharmonicity effects to some degree.

An example of a potential that has the correct qualitative form for the stretching of a diatomic is the Morse potential, shown in Fig. 5.3. It is constructed for a specific diatomic by knowing (or guessing) the

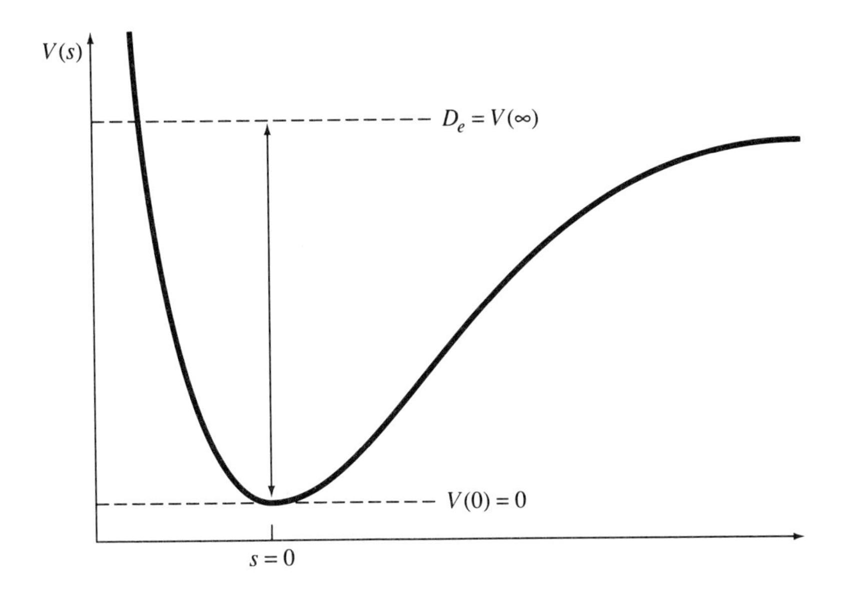


**FIGURE 5.3** The functional form of the Morse potential of Eq. (5-20).

dissociation energy, $D_e$, and choosing a parameter, $\alpha$, to yield the desired shape.

$$V^{Morse}(s) = D_e\left(1 - e^{-\alpha s}\right)^2 \tag{5-20}$$

where s is the coordinate corresponding to the displacement from equilibrium separation of the atoms. At s = 0, the potential is zero, and asymptotically it approaches $D_e$ as s approaches infinity. For the region $s < 0$, the potential rises very steeply, and so it has the proper qualitative form throughout. This potential can be compared with a harmonic potential by writing its power series expansion in s.

$$V^{Morse}(s) = D_e\left[\alpha^2 s^2 - \alpha^3 s^3 + \frac{7}{12}\alpha^4 s^4 - \frac{1}{4}\alpha^5 s^5 + \ldots\right] \tag{5-21}$$

The first term is the harmonic part. For $\alpha s < 1$, the typical situation, the terms diminish in size as the series continues, but even for $\alpha s \cong 1$, the terms diminish in importance, though *after* the cubic term. The cubic and quartic terms are the leading source of anharmonicity for near equilibrium displacements of a Morse oscillator, and in fact, it is very often the case that a cubic potential function is the most important contributor to anharmonicity effects of real molecular potentials.

To incorporate anharmonicity effects in a description of a diatomic, a third-order polynomial function is a starting point for the potential, and we may express this generally as

$$V(s) = \frac{1}{2}ks^2 + gs^3 \tag{5-22}$$

This is a starting point because higher order terms or other functional forms might be employed for still greater precision in representing a true potential. The cubic term, though, is often the most significant anharmonic element of a potential, as noted for the Morse potential. The cubic term (and any higher order terms) may be well treated as a perturbation of the harmonic oscillator. The first-order corrections for all states turn out to be zero. The second-order corrections arising from the cubic term $gs^3$ are

$$E_n^{(2)} = -\frac{7}{16}\frac{g^2\hbar^2}{\mu k^2} - \frac{15}{4}\frac{g^2\hbar^2}{\mu k^2}\left(n + \frac{1}{2}\right)^2 \tag{5-23}$$

A cubic potential term will usually have a negative sign; that is, $g < 0$ in Eq. (5-22), and this was the case for the Morse oscillator. Regardless of the sign of g, the associated anharmonicity corrections of Eq. (5-23) are a lowering of the energy of each vibrational state relative to the harmonic picture. Furthermore, the extent of lowering increases with n. The energy levels of a typical anharmonic oscillator tend to become more closely spaced with increasing energy as illustrated in Fig. 5.4.

## 5.4 SELECTION RULES AND SPECTRA

The vibrational-rotational energy levels of a diatomic are only part of the information needed to understand and interpret spectra. The other part is the information about what particular transitions among the states may take place in an experiment. From time-dependent perturbation theory

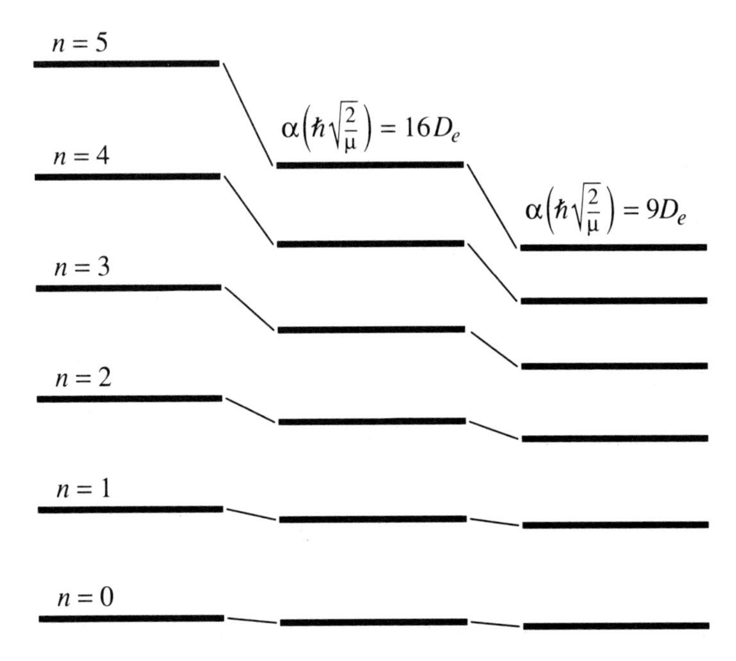


**FIGURE 5.4** A representative correlation diagram of several of the low-lying energy levels of a harmonic oscillator with the energy levels of an anharmonic oscillator. The anharmonic energy levels are those given by Eq. (5-23) assuming that the potential function is the power series expansion of a Morse potential truncated at the cubic term. The closer spacing of the levels with increasing vibrational quantum number is typical.

(Chap. 4), the allowed transitions are those among stationary states that are mixed by the perturbing Hamiltonian. If H' is a Hamiltonian corresponding to the interaction of the diatomic with electromagnetic radiation, then the stationary states (n,J) and (n',J') are mixed if the following is true.

$$\langle \Psi_{nJ} \mid H' \Psi_{n'J'} \rangle \neq 0$$

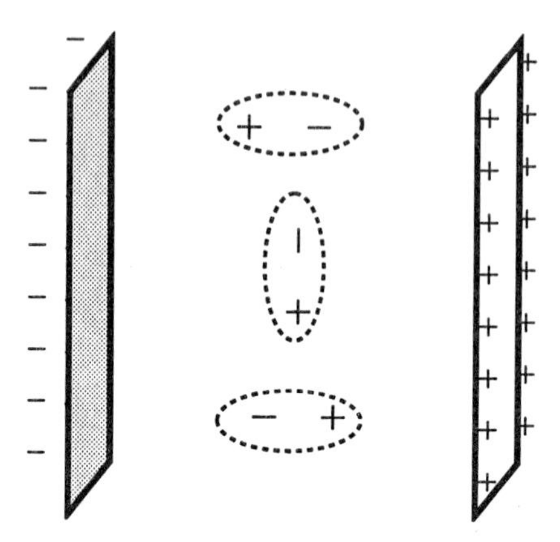

**FIGURE 5.5** An illustration of a dipole placed in a uniform electric field. The field develops between the two oppositely charged, parallel plates. A dipole moment is a property of a separated positive and negative charge, such as within the circled area. At the three orientations shown, there will be three different interaction energies because the positive end of the dipole is attracted toward the negatively charged plate and the negative end of the dipole is attracted toward the positively charged plate.

---

If this integral is nonzero, the transition from one state to the other is allowed (written $nJ \longleftrightarrow n'J'$ ). Otherwise, it is forbidden (written $nJ \not\longleftrightarrow n'J'$ ).

The perturbing Hamiltonian in a basic spectroscopic experiment is the interaction with the electric field of the radiation. A uniform electric field interacts via the dipole moment of a charge distribution; the energy of interaction is the negative of the dot product of the field, $\vec{E}$, and the dipole. Before considering the electric field of light, let us consider a static, uniform electric field. One way such a field can be generated is by oppositely charging two parallel plates as shown in Fig. 5.5. A simple charge distribution that would possess or give rise to a dipole moment is an arrangement of a positive point charge and a negative charge, as in Fig. 5.5. This simple illustration shows that the orientation of the charge distribution with respect to the field will affect the interaction energy. At $0^{\circ}$, the interaction is energetically favorable, which means it takes on a negative value. At $180^{\circ}$, the interaction energy should be the same size, but positive in sign. At $90^{\circ}$, the interaction is zero. This type of orientational dependence is mathematically expressed as a dot product of two vectors.

One vector points from one point charge to the other (the dipole moment vector) and the other vector points from one plate to the other (the electric field vector).

With radiation oscillating at a frequency $\omega_r$, the interaction Hamiltonian is

$$H' = -\vec{\mu} \bullet \vec{E} \cos(\omega_r t) \tag{5-24}$$

where $\vec{\mu}$ is the dipole moment vector. (Notice that $\mu$ is used both for dipoles and for the reduced mass of a two-body system, but the context should make it clear which $\mu$ is meant.) To treat electromagnetic radiation encountering molecules randomly oriented in space, we must assume an arbitrary or unspecified orientation. The orientation with respect to the molecular axis is the rotational angle, $\theta$, and since the dipole moment has a nonzero component only along the molecular axis, the dot product in Eq. (5-24) varies with the cosine of $\theta$. For a realistic picture of a molecule, we must allow for the dipole moment to vary with the geometry of the molecule. Were there a fixed distribution of charges, the dipole moment would be unchanging, but in a molecule, the vibrational excursions of the nuclei mean that the charge distribution is not fixed. For a diatomic molecule, then, the dipole moment is a function of the displacement, i.e., $\mu(s)$. Thus, the interaction Hamiltonian may be expressed as

$$H' = -|\vec{E}| \cos(\omega_r t)\, \mu(s) \cos\theta \tag{5-25}$$

The selection rules for dipole transitions will be determined by integrals with this operator.

So far, the wavefunctions for a diatomic have been labelled by the quantum numbers n and J. The quantum number M, which gives the z-component of the rotational angular momentum, has been suppressed. At this point, we need to make explicit use of the wavefunctions, and since these are products of radial and spherical harmonic functions, we will now include the M quantum number:

$$\psi_{nJM}(s,\theta,\phi) = S_n(s)\, Y_{JM}(\theta,\phi)$$

The general integral expression needed for determining the selection rules is

$$<\psi_{nJM}\,|H'\,\psi_{n'J'M'}> \tag{5-26a}$$

$$= -\,|\vec{E}|\cos(\omega_r t) < S_n(s)\,Y_{JM}(\theta,\phi)\,|\,\mu(s)\cos\theta\, S_{n'}(s)\,Y_{J'M'}(\theta,\phi) >$$

The integral factors into integrals over the radial and angular coordinates.

$$<\psi_{nJM}\,|H'\,\psi_{n'J'M'}> \tag{5-26b}$$

$$= -\,|\vec{E}|\cos(\omega_r t) < S_n(s)\,|\,\mu(s)\,S_{n'}(s) > \; < Y_{JM}(\theta,\phi)\,|\cos\theta\,Y_{J'M'}(\theta,\phi) >$$

It is the integral over the radial coordinate that involves the dipole moment function.

The integral in Eq. (5-26b) over $\theta$ and $\phi$ may be evaluated for specific spherical harmonic functions by explicit integration. It is also possible to obtain the following result that is general for any choice of J, M, J', and M'.

$$< Y_{JM}\,|\cos\theta\; Y_{J'M'} > \;=\; \delta_{MM'}\sqrt{\delta_{J,J'+1}\frac{J^2-M^2}{4J^2-1} + \delta_{J,J'-1}\frac{J'^2-M^2}{4J'^2-1}} \tag{5-27}$$

This expression is developed by using the recursion relations of the Legendre polynomials to replace $\cos\theta\, Y_{J'M'}$ by a sum of $Y_{J'+1,M'}$ and $Y_{J'-1,M'}$. Then, the orthonormality of the spherical harmonic functions yields the result. The immediate information from Eq. (5-27) is the selection rule that the J quantum number must change by one between the initial and final states of a dipole allowed transition. That is, the transition is allowed only if $J = J'-1$ or if $J = J'+1$, and that can be stated as $|J - J'| = 1$.

The integral over the radial coordinate in Eq. (5-26) can be evaluated in several ways. One particularly helpful approximation starts with a power series expansion of the dipole moment function, $\mu(s)$. The expansion is then truncated at some low order thought to be appropriate for the system at hand.

$$\mu(s) \;=\; \mu_e \;+\; s\left.\frac{d\mu}{ds}\right|_{s=0} \;+\; \frac{1}{2}s^2\left.\frac{d^2\mu}{ds^2}\right|_{s=0} \;+\; \ldots \tag{5-28}$$

$\mu_e = \mu(0)$ is the dipole moment when the molecule is at its equilibrium length ($s = 0$). Thus,

$$< S_n(s) \mid \mu(s) S_{n'}(s) > \;=\; \mu_e < S_n(s) \mid S_{n'}(s) > + \left.\frac{d\mu}{ds}\right|_{s=0} < S_n(s) \mid s\, S_{n'}(s) >$$

$$+ \frac{1}{2}\left.\frac{d^2\mu}{ds^2}\right|_{s=0} < S_n(s) \mid s^2 S_{n'}(s) > + \;\ldots$$

If the radial wavefunctions are those of the harmonic oscillator (that is, if the harmonic oscillator approximation of the potential is invoked), then explicit integration can be performed for any choice of n and n'. The result is the following expression (with m being used here for the reduced mass).

(5-29)

$$< S_n(s) \mid \mu(s) S_{n'}(s) > \;=\; \mu_e\, \delta_{nn'} + \left.\frac{d\mu}{ds}\right|_{s=0} \sqrt{\frac{\hbar}{2\sqrt{mk}}} \left( \sqrt{n}\; \delta_{n,n'+1} + \sqrt{n'}\; \delta_{n+1,n'} \right)$$

$$+ \frac{1}{2}\left.\frac{d^2\mu}{ds^2}\right|_{s=0} \frac{\hbar}{2\sqrt{mk}} \left( (2n+1)\, \delta_{nn'} + \sqrt{n(n-1)}\; \delta_{n,n'+2} + \sqrt{n'(n'-1)}\; \delta_{n+2,n'} \right) + \ldots$$

The interpretation of Eq. (5-29) is made term by term. The first term indicates that transitions are allowed wherein n and n' are the same, which means where the vibrational quantum number does not change, but only if there is a nonzero equilibrium dipole moment ($\mu_e$). The next term indicates that transitions will be allowed where the difference between n and n' is one, but only if there is a nonzero first derivative of the dipole moment function, $d\mu/ds$. Similarly, the next term depends on the second derivative of the dipole moment function. If it is not zero, transitions will be allowed for n = n' and for a change of 2 in the n quantum number.

Fig. 5.6 shows the dipole moment function of the hydrogen fluoride molecule. It shows a very typical feature of diatomic molecules, which is that the dipole moment function of the molecule is very nearly linear in the displacement coordinate in the vicinity of the equilibrium. Of course, at large hydrogen-fluorine separations, there is significant curvature. Approximating the dipole moment function by truncating the expansion in Eq. (5-28) after the term that is linear in s will, therefore, be a very good approximation for the lower energy vibrational states. When this approximation is made and when the stretching potential is taken to be strictly harmonic, then only the first two right hand side terms in Eq. (5-29)

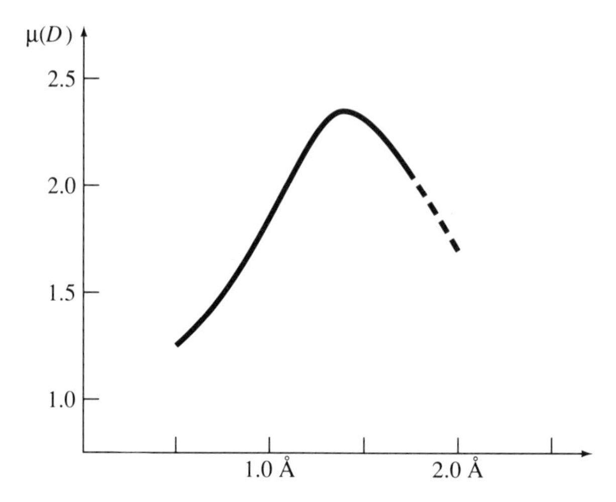


**FIGURE 5.6** The dipole moment function of the hydrogen fluoride molecule as a function of the separation distance between H and F in Å. This curve is based on spectroscopic data [R. N. Sileo and T. A. Cool, J. Chem. Phys. **65**, 117 (1976) and T. E. Gough, R. E. Miller and G. Scoles, Faraday. Discuss. Chem. Soc. **71**, 77 (1981)] with an extrapolation to the asymptotic value of zero at infinite separation. The equilibrium separation distance is at 0.92 Å, and in the vicinity of the equilibrium, the dipole moment curve is very nearly linear.

are carried. This is called a *doubly harmonic approximation*, and it gives as the selection rule that the vibrational quantum number can be unchanged or can change by one; however, there must be a permanent equilibrium dipole for the first case and a *changing* dipole for the second case. A molecule such as $N_2$ has a zero dipole moment because by symmetry it is nonpolar. That symmetry remains whether the molecule is stretched or compressed, and so it does not have a changing dipole (i.e., $d\mu/ds = 0$).

For a typical polar diatomic molecule, we conclude that electromagnetic radiation will induce changes in the states of the system wherein the vibrational quantum number changes by zero or one and the rotational quantum number changes by one. Though other transitions may be seen, these are always expected. Vibrational transitions are said to be *fundamental* if the vibrational quantum number changes from zero to one. They are said to be *overtones* if the vibrational quantum number changes from 0 to 2, 3, and so on.

The selection rules are crucial information for working from a laboratory spectrum back to a bond length or harmonic force constant. But to do that requires that we change our thinking somewhat. So far, we have learned how to *predict* a spectrum given the molecule's potential, V(s). But the stretching potentials of diatomics are not known *a priori*. In fact, they are the very information that is to be extracted from experiment, not the other way around. Analyzing and interpreting spectra and extracting molecular information require matching predictions, based on quantum mechanics, with laboratory measurements.

The result of a spectroscopic experiment is a set of transition frequencies, as well as the intensities or strengths of the lines. As an example of the information that is available from an experiment, Fig. 5.7 shows an idealization of the high-resolution infrared spectrum of the hydrogen fluoride molecule. With modern instrumental techniques, the frequencies of the lines in this spectrum can be measured with a precision of about 0.01 $cm^{-1}$ and better. Each of the spikes or lines is at a frequency that corresponds to a particular transition between quantum states of HF. Since the frequency of radiation is proportional to the energy of the photons, or the energy absorbed by a molecule, then knowing the frequency of a transition is equivalent to knowing the energy of the transition. In fact, wavenumbers or $cm^{-1}$ serve as units of measure for both frequency and energy (see Appendix V).

The length or height of each spike or transition line is related to the intensity of that transition, which is to say that the depletion of incident radiation at a given frequency depends on the likelihood that the sample will absorb at that frequency. The intensities and frequencies are the primary data from the spectrum. The pattern of the lines appears to be regular or organized, but it needs to be deciphered or analyzed in order to make use of the data. We will now consider the analysis of the frequencies of the lines and then the intensities.

The energy of a transition is the difference between the energies of an initial state and a final state. Therefore, instead of an expression for the energies of the vibrational-rotational states of a diatomic, we need an expression for energy differences. This can be accomplished by combining the information in an energy level expression with that in the selection rules. The result will be a transition energy expression, and it will contain

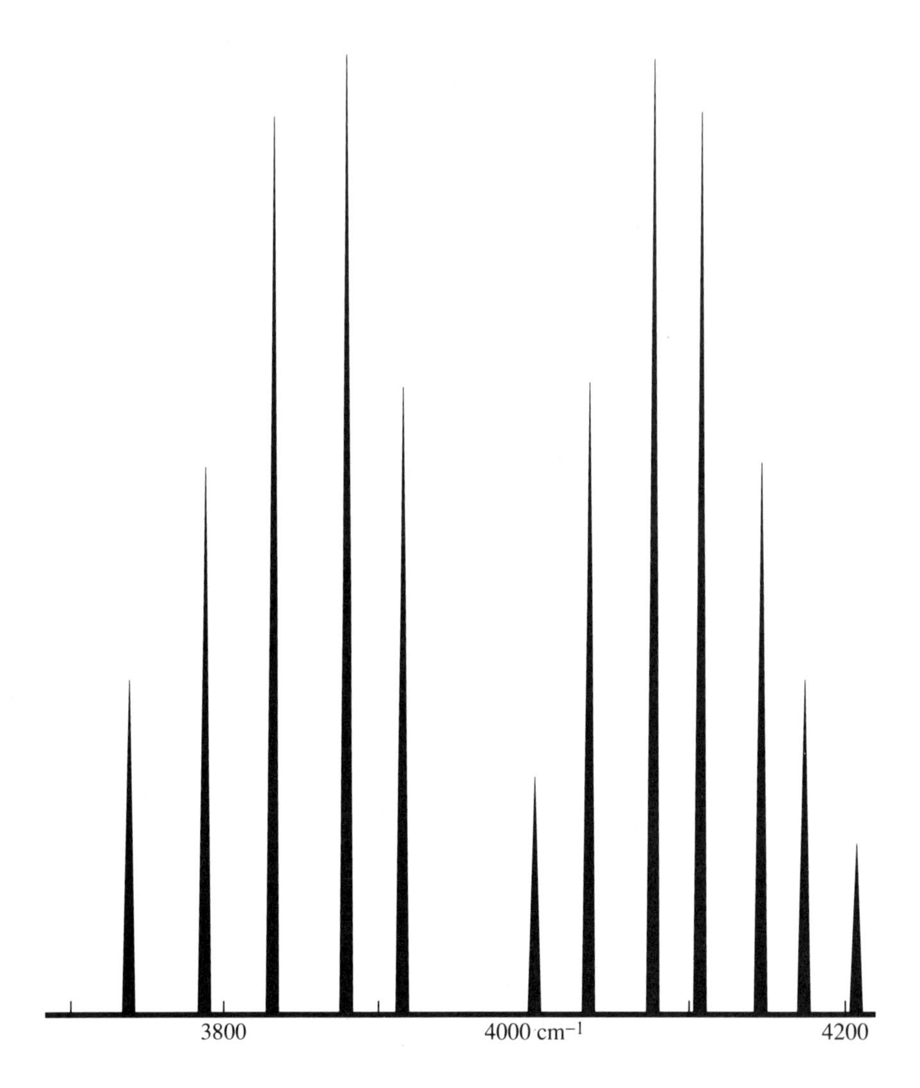


**FIGURE 5.7** An idealized high-resolution spectrum (i.e., a spectrum re-drawn to suppress instrumental noise, etc.) of a low-density gas phase sample of hydrogen fluoride. The frequencies of the peaks are the transition frequencies, and the relative heights of the peaks correspond to the intensities. This drawing shows the characteristic form of an infrared spectrum of a diatomic molecule: two sets of lines, called branches, with diminishing peak height to the left and to the right. The branch on the left, at lower transition frequencies, is the P-branch whereas the branch on the right is the R-branch. The separation between the two branches is roughly twice the separation between lines within the branches.

the information about where in the spectrum each of the allowed transitions should be found.

To start with the simplest level of treatment, let us use the energy level expression of Eq. (5-13) and the doubly harmonic selection rules. These rules say that upon going from some initial n,J state to another state by absorption of electromagnetic radiation,

i. J must increase or decrease by one, *and*

ii. n must increase by one.

This presents two situations, one where J increases and one where J decreases. It is easiest to obtain two transition energy expressions, one for each case. For the first situation, an energy difference expression is obtained by subtracting the initial state's energy, expressed according to Eq. (5-13), from the final state's energy:

$$E_{n+1,J+1} = (n + \frac{3}{2})\hbar\omega + \frac{\hbar^2 (J+1)(J+2)}{2\mu r_e^2}$$

minus:

$$E_{n,J} = (n + \frac{1}{2})\hbar\omega + \frac{\hbar^2 J(J+1)}{2\mu r_e^2}$$

$$\Delta E_{n,J\rightarrow n+1,J+1} = \hbar\omega + \frac{\hbar^2}{2\mu r_e^2}(2J + 2) \quad (5\text{-}30)$$

This is an energy difference for the allowed transition where the initial state's J quantum number increases by one.

The other situation is that where the J quantum number decreases by one. This requires a separate energy difference expression:

$$E_{n+1,J-1} = (n + \frac{3}{2})\hbar\omega + \frac{\hbar^2 (J-1)J}{2\mu r_e^2}$$

minus:

$$E_{n,J} = (n + \frac{1}{2})\hbar\omega + \frac{\hbar^2 J(J+1)}{2\mu r_e^2}$$

$$\Delta E_{n,J\rightarrow n+1,J-1} = \hbar\omega - \frac{\hbar^2}{2\mu r_e^2}2J \quad (5\text{-}31)$$

These two ΔE expressions should be regarded as predictions of the positions of transition lines that may be observed in the spectrum, at least according to the particular level of treatment that was used for the state energies.

In both Eqs. (5-30) and (5-31), the same set of constants shows up in the second term. Collecting them into one constant, B, called the *rotational constant*, makes it simpler to write the expressions. Also, if the energy, E, the vibrational frequency, ω, and the rotational constant, B, are all expressed in $cm^{-1}$, then these expressions become quite concise.

$$\Delta E_{n,J \to n+1,J+1} \, (cm^{-1}) \;=\; \omega \;+\; 2B\,(J+1) \tag{5-32}$$

$$\Delta E_{n,J \to n+1,J-1} \, (cm^{-1}) \;=\; \omega \;-\; 2BJ \tag{5-33}$$

B is much smaller than ω, and so the transitions in the spectrum will all be clustered in the vicinity of the frequency ω. This is the first feature to identify in the spectrum. For instance, in Fig. 5.7, the lines appear centered about 3960 $cm^{-1}$, and so that serves as a fair idea of the value of ω.

The next thing to do with the spectrum is to *assign* individual lines. Assignment of a spectrum means that the transition lines have been associated with particular initial and final states. For a vibrating-rotating diatomic, this means identifying the n,J quantum numbers for the initial state and the n',J' quantum numbers for the final state for each line. An ordered list of transition energies, generated with Eqs. (5-32) and (5-33), reveals how to make these assignments. We start with Eq. (5-33) because it gives frequencies that are all below those given by Eq. (5-32); it must give the leftmost line seen in the spectrum. We pick out a value of J and work up in frequency until reaching J = 1, which is where the highest frequency possible with Eq. (5-33) is reached. Then, we use Eq. (5-32) and start with J = 0, the lowest frequency that this equation can give, and then work up.

| J (initial) | → J' (final) | ΔE | Δ(ΔE) |
|---|---|---|---|
| 4 | 3 | ω − 8B | |
| | | | 2B |
| 3 | 2 | ω − 6B | |
| | | | 2B |
| 2 | 1 | ω − 4B | |
| | | | 2B |
| 1 | 0 | ω − 2B | |
| | | | **4B** |
| 0 | 1 | ω + 2B | |
| | | | 2B |
| 1 | 2 | ω + 4B | |
| | | | 2B |
| 2 | 3 | ω + 6B | |
| | | | 2B |
| 3 | 4 | ω + 8B | |

This table shows that there will be a number of transition frequencies in the vicinity of $\omega$, but offset by some multiple of 2B.

$\Delta(\Delta E)$ are the differences between transition energies. They tell how far apart adjacent spectral lines are. The pattern here is one of lines separated by 2B, except at one crucial point where the separation is twice as much, 4B. Examining Fig. 5.7 shows that in the middle of the cluster of lines, there is a point where the separation is about twice as big as otherwise seen. Thus, the line to the left (lower frequency) of this separation is assigned to be the J = 1 to J' = 0 transition (with n changing by one, as well). The line to the right, then, corresponds to an initial state of J = 0 and a final state where J' = 1. The remaining lines follow in the sequence in the table.

A confirmation of the assignment comes from the relative intensities. Notice that the strengths of the transitions, or the heights of the lines in the spectrum, diminish to the right and to the left. That is, the transitions with the highest initial J quantum number are weakest. The primary reason for this has to do with the *population* of the different quantum states, or the number of molecules in each of the states. The *Maxwell-Boltzmann distribution law*, which is stated without development or proof, is that for a sample of N molecules in thermal equilibrium at a temperature T, where the molecules may exist in stationary states with energies $E_i$ (for each of the $i^{th}$ states), the number of molecules in each state, $N_i$, is

$$N_i = N \frac{e^{-E_i / kT}}{\sum_j e^{-E_j / kT}} \tag{5-34}$$

where k is the *Boltzmann constant*. The denominator in Eq. (5-34) is called the partition function, and in this expression, it serves to ensure that $\Sigma N_i$ = N. The Maxwell-Boltzmann law says that the number of molecules in a state will diminish exponentially with the energy of the state for some given temperature.

If we use the vibrational-rotational energy level expression of Eq. (5-13), the population of some state with quantum numbers n, J, and $M_J$, relative to the population of the state with the same vibrational quantum number but with J = 0, is,

$$\frac{N_{n,J,M_J}}{N_{n,0,0}} = \frac{e^{-E_{nJ}/kT}}{e^{-E_{n0}/kT}} = e^{-\hbar^2 J(J+1)/(2\mu r_e^2 kT)}$$

So, as J becomes larger, the population diminishes relative to the population of the lowest J state (J = 0). We have already realized that transitions may be observed with any initial J quantum number, but because there are different numbers of molecules in each J state, the number of transitions originating from a given J state must follow proportionately.

One further point is that the transition lines do not distinguish the $M_J$ quantum number. For any initial nJ energy level, transitions may occur from each and every one of the $M_J$ states, and there are (2J+1) of these. This means that a transition strength will depend on the population of an nJ energy level, rather than on the population of a single state. The population of an energy level must be the sum over the populations of all the states of that energy. But according to Eq. (5-34), the populations of states of the same energy are identical. So, the population of the level is simply the population of any one state times the number of degenerate states, which is a factor called the degeneracy. For vibrational-rotational states, this factor, often designated $g_J$, is simply (2J+1). Therefore, the relative intensities, I, of the lines in a high-resolution vibrational absorption band, as in Fig. 5.7, will depend on the initial state's J quantum number,

$$I_J \propto (2J+1)\, e^{-\chi J(J+1)} \quad \text{where } \chi = \hbar^2 / (2\mu r_e^2 k T)$$

Fig. 5.8 is a plot of this function for different values of $\chi$, treating J as a continuous variable. Notice that a curve drawn through the tops of the transition lines in Fig. 5.7 would have the form of a curve in Fig. 5.8. Indeed, at low resolution, where the individual lines overlap or are not resolved, the branches of an infrared absorption spectrum may resemble the curves in Fig. 5.8.

The analysis so far provides a basis for assigning lines in the infrared spectrum of a diatomic. Since quantum mechanical analysis provides energy level expressions subject to the level of approximation chosen, then in principle, highly detailed information may be extracted from spectral data. The J dependence in an energy level expression is always in terms of

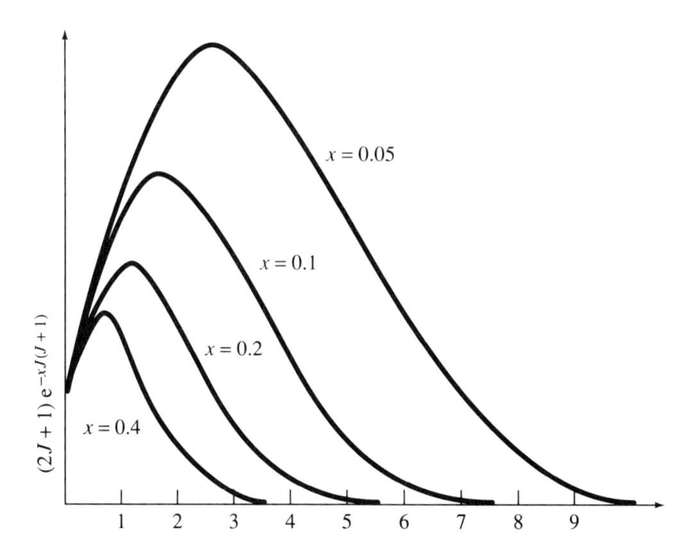


**FIGURE 5.8** A plot of the relative transition intensities of the rotational lines in a diatomic molecule's vibrational absorption band. The curves are plots of the quantity (2J + 1) exp[$-\chi$ J(J+1)] but with J treated as a continuous variable. The numerical values of $\chi$ are shown next to each curve. Notice that as the temperature increase, $\chi$ becomes smaller.

the quantity J(J+1), and the n dependence is always in terms of (n+1/2). An overview of Eqs. (5-13,17,20,23), the various energy level expressions, reveals that they all take the form of an expansion of the energy in terms of polynomials* in (n+1/2) and J(J+1). That is, the energy of an nJ level has the following general form.

$$E_{nJ} = \sum_{i=0}^{\infty} \sum_{k=0}^{\infty} c_{ik} \left(n + \frac{1}{2}\right)^i [J(J+1)]^k \tag{5-36}$$

The various approximations that have been discussed lead to certain truncations of this expansion and to specific values of the constants, $c_{ik}$. For instance, we may say that Eq. (5-13) is of the form of Eq. (5-36) with $c_{10} = \hbar\omega$, $c_{01} = \hbar^2/(2\mu r_e^2)$, and all other $c_{ik}$'s equal to zero.

* It is helpful to regard (n+1/2) as a single variable, e.g., x, and J(J+1) as another independent variable, e.g., y. The polynomial expansion is then just a standard type, $E(x,y) = c_{00} + c_{10}x + c_{01}y + c_{11}xy + c_{20}x^2 + c_{02}y^2 + \ldots$ .

Transition energy expressions may be developed for any desired truncation of Eq. (5-36) by means of the differencing procedures in the prior sections. These will give $\Delta E$ for the transition lines in terms of the $c_{ik}$'s. Analysis of spectra may then be carried out by first assigning lines, and then by fitting the measured transition frequencies (energies) to the $\Delta E$ expression being employed (see Appendix I). Notice how this has taken the whole approach one step away from the details of the quantum mechanical analysis: The measured transition energies or frequencies must fit a simple polynomial expression, Eq. (5-35), and the main task is to find the right $c_{ik}$'s. At that stage, we look back at the quantum mechanical analysis and relate the $c_{ik}$'s to various kinds of molecular information, such as the bond length and the stretching force constant.

There is well-established terminology and notation for the most important $c_{ik}$'s of Eq. (5-35), and some of this has been introduced already. The $c_{ik}$'s are spectroscopic constants; that is, they are values deduced or extracted from measurement of spectra. The names and designations for the most important ones are the following. The corresponding $c_{ik}$'s are shown in brackets; in some cases, the conventional definition of the spectroscopic constants introduces a negative sign, and so the values may turn out to correspond to $-c_{ik}$'s.

| | |
|---|---|
| $\omega_e$ | harmonic (equilibrium) vibrational frequency [$c_{10}$] |
| $\omega_e\chi_e$ | equilibrium anharmonicity; this product is considered to be one value [$-c_{20}$] |
| $\omega_e y_e$ | equilibrium second anharmonicity constant; this product is considered to be one value [$c_{30}$] |
| $B_e$ | equilibrium rotational constant [$c_{01}$] |
| $D_e$ | equilibrium centrifugal distortion constant [$-c_{02}$] |
| $H_e$ | equilibrium second centrifugal distortion constant [$c_{03}$] |
| $\alpha_e$ | vibration-rotation coupling constant [$-c_{11}$] |
| $\gamma_e$ | [$c_{21}$] |
| $\beta_e$ | [$c_{12}$] |

The most common system of units for reporting infrared spectroscopic constants is that of wavenumbers or $cm^{-1}$ (see Appendix V). A tilde is sometimes used to emphasize that these are the units. So, the vibrational frequency in wavenumbers, $\tilde{\omega}_e$, is related to $\omega_e$ in $sec^{-1}$ by a factor of $2\pi c$. (Of course, as discussed in Appendix V, we may freely state $\omega$ and other constants in $cm^{-1}$, $sec^{-1}$, ergs, and other units, and as long as the units are clear from the context, a special symbol is not really necessary.) Wavenumbers are suitable as an energy scale, and consequently, we may write an energy expression such that the energy and every constant are taken to be in wavenumbers. So, with the named spectroscopic constants in place of the $c_{ik}$'s in Eq. (5-35), we have

$$\tilde{E}_{nJ} = \tilde{c}_{00} + \tilde{\omega}_e (n + \frac{1}{2}) - \tilde{\omega}_e\tilde{\chi}_e (n + \frac{1}{2})^2 + \tilde{\omega}_e\tilde{y}_e (n + \frac{1}{2})^3 + \tilde{B}_e J(J+1)$$

$$- \tilde{\alpha}_e J(J+1) (n + \frac{1}{2}) + \tilde{\gamma}_e J(J+1) (n + \frac{1}{2})^2 - \tilde{D}_e [J(J+1)]^2$$

$$- \tilde{\beta}_e [J(J+1)]^2 (n + \frac{1}{2}) + \tilde{H}_e [J(J+1)]^3 + \ldots$$

The value $\tilde{c}_{00}$ cannot be obtained by measurement of transition frequencies because it will not show up in an energy difference ($\Delta E$) equation. It contributes, usually in a small way, to the zero point energy of the system. Table 5.1 lists spectroscopic constants of a number of diatomics.

## 5.5 ROTATIONAL SPECTROSCOPY

Analysis of the microwave or radiofrequency spectra of diatomics is usually simpler than the analysis of their vibrational spectra. The energy of microwave radiation is so much less than the photon energies of infrared radiation that the vibrational quantum number does not change in the course of a transition. Thus, we need to consider energy differences having to do with rotation but not vibration. Of course, with no change in n, there is the requirement that the molecule have a permanent dipole moment for any transitions to be allowed. Then, the selection rule is simply that the allowed transitions are those for which J increases by one. So, instead of an infrared band of transitions arising from the populated rotational (J) levels, a single transition line corresponding to $J = 0 \rightarrow 1$ is usually observed in the microwave or radiofrequency spectrum of a diatomic. The appropriate transition energy expression is obtained by taking differences using the energy level expression for the vibrational-rotational states of a diatomic;

TABLE 5.1 Spectroscopic Constants (in $cm^{-1}$) of Certain Diatomic Molecules. [a]

| | $\omega_e$ | $B_e$ | $\omega_e\chi_e$ | $\alpha_e$ |
|---|---|---|---|---|
| BN | 1514.6 | 1.666 | 12.3 | 0.025 |
| BO | 1885.4 | 1.7803 | 11.77 | 0.0165 |
| BaO | 669.8 | 0.3126 | 2.05 | 0.0014 |
| BeO | 1487.3 | 1.6510 | 11.83 | 0.0190 |
| CH | 2861.6 | 14.457 | 64.3 | 0.534 |
| CD | 2101.0 | 7.808 | 34.7 | 0.212 |
| CN | 2068.7 | 1.8996 | 13.14 | 0.0174 |
| CO | 2170.2 | 1.9313 | 13.46 | 0.0175 |
| CaH | 1299 | 4.2778 | 19.5 | 0.0963 |
| HBr | 2649.7 | 8.473 | 45.21 | 0.226 |
| HCl | 2989.7 | 10.59 | 52.05 | 0.3019 |
| HF | 4138.5 | 20.939 | 90.07 | 0.770 |
| DF | 2998.3 | 11.007 | 45.71 | 0.293 |
| HgH | 1387.1 | 5.549 | 83.01 | 0.312 |
| LiH | 1405.6 | 7.5131 | 23.20 | 0.2132 |
| MgH | 1495.7 | 5.818 | 31.5 | 0.1668 |
| MgO | 785.1 | 0.5743 | 5.1 | 0.0050 |
| NO | 1904 | 1.7046 | 13.97 | 0.0178 |
| OH | 3735.2 | 18.871 | 82.81 | 0.714 |
| OD | 2720.9 | 10.01 | 44.2 | 0.29 |
| SiN | 1151.7 | 0.7310 | 6.560 | 0.0057 |
| SiO | 1242.0 | 0.7263 | 6.047 | 0.0049 |

[a] These values have been collected in Ref. 6 in the bibliography for this chapter. The values are for the most abundant isotopes, except for the deuterated molecules CD, DF, and OD.

however, because the n quantum number is unchanging, and since $\Delta J = +1$, the transition energy expression is rather simple.

A $J = 0$ to $J = 1$ transition will be observed at a frequency of about 2B, whereas the next transition, $J = 1$ to $J = 2$, will be at 4B, which is twice the frequency. In the microwave and radiofrequency region of the electromagnetic spectrum, a doubling of the frequency may require instrumental alterations. So, only one transition is usually seen for a particular equipment setup. (Of course, for heavy diatomic molecules, the rotational constant, B, is small, and then it may turn out that several transitions, such as $J = 4$ to 5, $J = 5$ to 6, and $J = 6$ to 7, are close enough in frequency as to be observable in one scan.)

It is characteristic of the technology of microwave spectroscopy that frequencies are measurable to very high precision. Until the introduction of infrared lasers, microwave spectroscopy outran vibrational spectroscopy in the precision and accuracy of spectral measurements. The main information obtained from a microwave spectrum is the rotational constant, and so the precision available with this type of experiment means that precise values of the rotational constant are obtained. This, in turn, implies that very precise values of the bond length of a diatomic can be deduced from a microwave spectrum.

The rotation of a rigid, linear triatomic or polyatomic molecule is mechanically equivalent to the rotation of a rigid diatomic. All are the rotations of an infinitesimally thin rod with two or more point masses attached. So, the basic analysis needed to understand the rotational spectroscopy of linear polyatomic molecules follows that of diatomics at the outset. The difference amounts to a generalization of the moment of inertia when there are more than two atoms. From classical mechanics it is known that the moment of inertia, I, about the center of mass of a linear arrangement of point masses is

$$I = \sum_{i<j} m_i m_j r_{ij}^2 \Big/ \sum_i m_i \qquad \text{(5-37)}$$

where $r_{ij}$ is the distance between the i and j atoms, $m_i$ is the mass of the $i^{th}$ atom, and the sum in the numerator is over all pairs of atoms in the molecule. The denominator, of course, equals the total mass of the molecule. For a diatomic molecule, Eq. (5-37) reduces to a single term,

$m_1 m_2 r_{12}^2 / (m_1 + m_2)$, and this is the same as $\mu r_e^2$, taking $r_e$ to be the same thing as $r_{12}$.

The rotational Schrödinger equation is no different than that for a diatomic because of the mechanical equivalence.

$$\frac{\hat{L}^2}{2I} \psi(\theta,\phi) = E \psi(\theta,\phi) \tag{5-38}$$

is the generalization of the two-body rotational Schrödinger equation and it comes about merely by replacing $\mu r^2$ by I. Therefore, the wavefunctions for a rigid, rotating diatomic or linear polyatomic must be the spherical harmonic functions because they are the eigenfunctions of $\hat{L}^2$. The eigenenergies are

$$E_J = J(J+1)\frac{\hbar^2}{2I} \tag{5-39}$$

where J is the rotational quantum number.

The rotational constant is inversely related to the moment of inertia, and the moment of inertia is a function of the bond lengths. For a diatomic molecule, measurement of the rotational constant implies determination of the bond length. But for a linear triatomic molecule, there are two unknowns in the problem, the two bond lengths. One value, the value of B, is not sufficient to find two bond lengths.

For a linear polyatomic of N + 1 atoms, there are N bond lengths to be found in order to know the structure of the molecule (assuming there is a basis for already knowing that the molecule in question is linear). Finding the bond lengths is basically a problem of establishing N equations for the N unknowns. The N equations are established from measured rotational constants for isotopically substituted forms of the molecule. Isotopic substitution is a change in an atomic mass, and with that comes a change in the molecule's moment of inertia. However, we assume that the bond lengths are unchanged because the chemical bonding is surely unaffected by the numbers of neutrons in the nuclei. (Actually, this assumption is valid only in regard to equilibrium structures. For the on-average length of a bond where there is zero-point vibrational motion, isotopic substitution may influence the vibrational averaging slightly and thereby have an effect on the bond length. But this can be taken into account in a more detailed analysis.)

For each isotopic form of a molecule with an experimentally determined B, and hence I, we may write Eq. (5-37) with the bond lengths as the only unknowns. So, with N isotopic forms, we have N equations from which to find the N unknowns. A standard example is HCN. Microwave transition frequencies for the J = 0 to J = 1 transitions were found to be 88,631.6 MHz for $H^{12}C^{14}N$ and 72,414.6 MHz for $D^{12}C^{14}N$ [W. Gordy, *Phys. Rev.* **101**, 599 (1956)]. The B rotational constants are one-half of these values, and if they are converted to moments of inertia, we have that I(HCN) = $18.937 \times 10^{-40}$ g-cm$^2$ and I(DCN) = $23.178 \times 10^{-40}$ g-cm$^2$. [See Ref. 7 (p. 187) in the bibliography for this chapter.] The two equations to solve are

$$I(\mathrm{HCN}) =$$

$$\left( m_H m_C r_{HC}^2 + m_C m_N r_{CN}^2 + m_H m_N (r_{HC} + r_{CN})^2 \right) / \left( m_H + m_C + m_N \right)$$

$$I(\mathrm{DCN}) =$$

$$\left( m_D m_C r_{HC}^2 + m_C m_N r_{CN}^2 + m_D m_N (r_{HC} + r_{CN})^2 \right) / \left( m_D + m_C + m_N \right)$$

Using the atomic masses from Appendix IV, the first equation becomes

$$\frac{18.937 \times 10^{-40}\ \text{g-cm}^2 \times 27.0109}{1.66057 \times 10^{-24}\ \text{g/amu} \times 10^{-16}\ \text{cm}^2/\text{Å}^2} = 308.0308\ \text{Å}^2$$

$$= 12.0939\, r_{HC}^2 + 168.0369\, r_{CN}^2 + 14.1126 \left( r_{HC}^2 + 2\, r_{HC}\, r_{CN} + r_{CN}^2 \right)$$

27.0109 is the total mass of HCN in amu. The corresponding equation for DCN is

$$391.0607\ \text{Å}2 = 24.1692\, r_{HC}^2 + 168.0369\, r_{CN}^2$$

$$+ 28.2036 \left( r_{HC}^2 + 2\, r_{HC}\, r_{CN} + r_{CN}^2 \right)$$

Solution of these two simultaneous quadratic equations will produce the H-C and C-N bond lengths in Å. The values are 1.066 Å and 1.156 Å, respectively.* These two values are termed the *substitution structure*

* These lengths have been rounded, and so, 307.9746 instead of 308.0308 is obtained for HCN, and 391.2677 instead of 391.0607 for DCN.

of HCN since they have been obtained on the basis of isotopic substitution. The procedure can be used to determine the substitution structure of any linear polyatomic molecule, though solution of the N quadratic equations can be tedious enough to benefit from utilizing a computer program.

For nonlinear molecules, three rotational constants are used in the energy level expressions. They are associated with rotations about three orthogonal axes in the nonlinear molecule called *principal axes*. One or more of the rotational constants can be measured from a microwave spectrum, and isotopic substitution studies can be employed to extract geometrical parameters (i.e., bond lengths and angles). Of course, the analysis, so far, has assumed that the molecules are rigid, but this is only an approximation. The molecules are vibrating, and so the rotational constants represent a vibrational average of structural parameters. Also, just as in the case of diatomic molecules, molecular rotation may give rise to centrifugal distortion effects.

It is an interesting fact that rotational transitions of molecules can be observed for molecules in interstellar space. In this case, the transitions are from higher energy states to lower energy states, and so they involve emission of a photon instead of absorption. The selection rules are the same, and rotational emission spectra match absorption spectra. In large interstellar clouds, molecules may be rotationally excited by collisions with other atoms or molecules and by other processes. They may then emit photons as they de-excite to a lower state. Transitions that occur at radiofrequencies are observed on Earth by huge antennas called radiotelescopes. Molecules are readily "fingerprinted" by their rotational transition frequencies, and so radiotelescopes can be employed to detect the existence of molecules in space. That is, by operating a radiotelescope at the specific transition frequency of a particular molecule, one can determine if those molecules exist in the region of space on which the telescope is focused. In fact, the signal strength can be used to determine the relative abundance of molecules, and maps of various interstellar clouds have been constructed that show the varying relative amounts of certain species. Fig. 5.9 shows the telescope signal obtained over a very narrow frequency range, and the emission lines that were detected are identified with certain specific molecules. Clearly, radiotelescopes are an application of molecular spectroscopy on a very large scale.

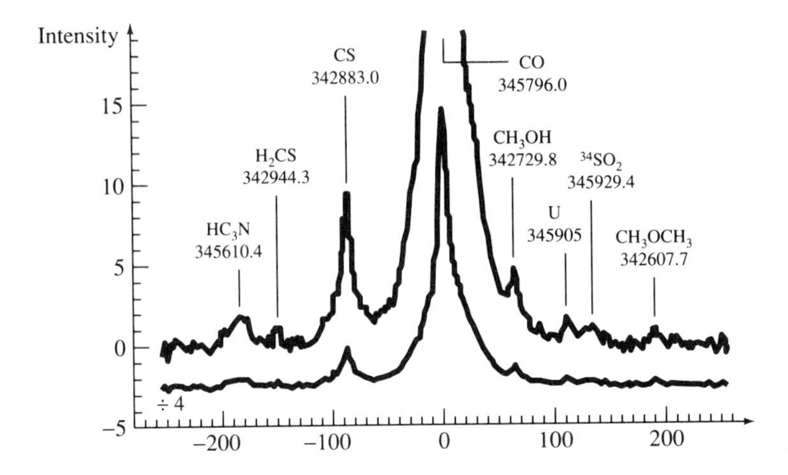


**FIGURE 5.9** Emission line spectrum of Orion A showing the existence in space of CO, CS, and several polyatomic molecules. Transition frequencies are given in MHz, and the spectrum is shown twice, once at full scale and once at 1/4 the vertical scale. [Reprinted with permission from the report by P. R. Jewell, J. M. Hollis, F. J. Lovas and L. E. Snyder, *Astrophys. J. Suppl. Series* **70**, 833 (1989).]

## 5.6 THE HARMONIC PICTURE OF POLYATOMIC VIBRATIONS

In low-resolution infrared spectroscopy of diatomic molecules, the rotational fine structure is lost, and some feature in a spectrum assigned to be a fundamental transition is but a single peak. The frequency of that peak is taken to be the vibrational frequency in the absence of any more precise experiments, and that is the extent of the information obtained. So, this low-resolution information corresponds mostly with a nonrotating picture or else a rotationally averaged picture of the molecule's dynamics; to the extent that we can analyze the data, we need only consider pure vibration. To understand the internal dynamics of polyatomic molecules, it is helpful to start with a "low-resolution" analysis. This means neglecting rotation, or presuming the molecules to be nonrotating. Further on and with heightened sophistication, we may consider a molecule to be both vibrating and rotating, as it would be in a gas phase sample.

The pure vibrations of a polyatomic molecule may be quite complicated. It is convenient to think first about the potential energy for vibrational motions in order to understand their nature. The potential energy will be a function of the atomic positions. For a molecule of N atoms, there are 3N atomic degrees of freedom, but only $3N-6$ ($3N-5$ for linear molecules) are left after removing the degrees of freedom for molecular translation and rotation. Thus, there are $3N-6$ (or $3N-5$) coordinates that describe the structure of the molecule but that do not give the molecule's position and orientation in space. These $3N-6$ (or $3N-5$) coordinates are called *internal coordinates*, and most often the bond lengths and bond angles comprise a suitable, though not unique, set. For instance, the internal structure of a water molecule can be specified by the two O-H bond lengths and the H-O-H bond angle. These constitute a set of three internal coordinates, and of course, $3N-6=3$ for water.

A *force field* is any potential for the vibrations of a molecule expressed in terms of some chosen set of internal coordinates. In principle, we arrive at complete understanding of the vibrations of a molecule if we know the force field precisely. Thus, we think of vibrational information in relation to the force field, and this is analogous to thinking about the vibrations of a diatomic in relation to the functional form of the stretching potential, V(x). Approximate force fields may be constructed (sometimes guessed) in several different ways, and they may be used for deducing or computing vibrational information. Laboratory measurement of vibrational frequencies provides the ultimate test of such computed information.

The simplest force field for a molecule is one that is harmonic. This means that the potential energy has only linear and quadratic terms involving the $3N-6$ coordinates. Some of these terms may be cross-terms, e.g., a product of two coordinates $r_1 r_2$. As discussed in Chap. 2, a potential that is harmonic in all coordinates may be written so that there are no cross-terms if the original coordinates are transformed to the normal coordinates. We will designate normal coordinates as $\{q_1, q_2, q_3, \ldots\}$, and the classical Hamiltonian in terms of normal coordinates is

$$H = \frac{1}{2} \sum_{i=1}^{3N-6} \left( \dot{q}_i^2 + \omega_i^2 q_i^2 \right) \tag{5-40}$$

This is obviously a separable problem, and as we have already considered in Chap. 2, it is equivalent to a problem of $3N-6$ independent harmonic oscillators. The vibrational frequencies of the oscillators are the $\omega_i$'s. The Schrödinger equation that develops from the quantum mechanical form of this Hamiltonian is also separable. The energy level expression comes from the sum of the eigenenergies of the separated harmonic oscillators or modes. For each, there is a quantum number, $n_i$.

$$E_{n_1 n_2 n_3 \ldots} = \sum_{i=1}^{3N-6} \left( n_i + \frac{1}{2} \right) \hbar \omega_i \tag{5-41}$$

The quantum numbers may take on values of 0, 1, 2, ....

A vibrational state of the polyatomic is specified by a set of values for the $3N-6$ quantum numbers in Eq. (5-41). The lowest energy state is the state that has all the quantum numbers equal to zero. The energy of this state, which is $\hbar \Sigma \omega_i / 2$, is the zero-point vibrational energy. Excited vibrational states are the infinite number of states for which any or all of the quantum numbers are not zero. As shown for the water molecule in Fig. 5.10, the energy levels for even a small polyatomic molecule become increasingly numerous at higher and higher energies above the ground state. That is, with the states arranged by their energies, the number of states found within some small energy increment is in an overall way increasing with energy. With states arranged by their energies, the counting of the number of states per some unit energy value is often referred to as the *density of states*. The density of vibrational states of a polyatomic molecule is a function of the energy that is increasing with energy.

From our detailed examination of diatomic vibrational motion, we know that the harmonic picture is an approximation with notable limitations. It is an approximation that is at its best for small amplitude displacements, and that means for low energy states. If we restrict attention to the low-energy states, then there is important qualitative information that goes with the harmonic picture. This qualitative information is the nature of the normal modes of vibration. The coordinate transformation from atomic displacement coordinates to normal coordinates is a transformation that is equally valid for a classical or a quantum mechanical picture, and so we may use classical notions about normal mode vibrations to understand the forms of molecular normal mode vibrations.

Let us use the carbon dioxide molecule as an example. It should have

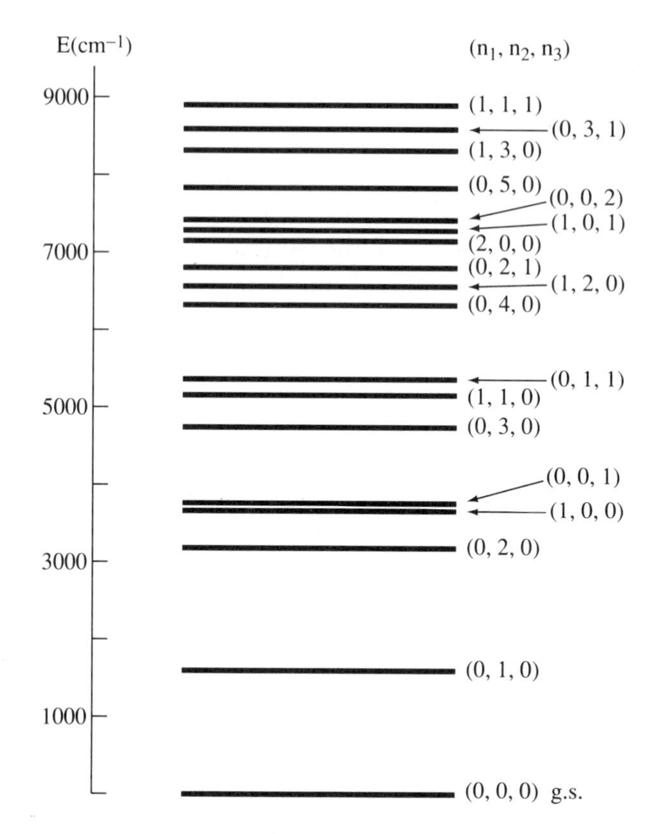


**FIGURE 5.10** A representation of the low-lying vibrational state energy levels of the water molecule in a harmonic picture showing the increasing number of states per energy interval at higher energies.

---

3N – 5 or four modes of vibration. We may represent each mode by the "direction" of the associated normal coordinate, and that means the directions in which each of the atoms move in the course of the vibration. Recall that a normal mode of vibration is the simplest motion of a system of particles, and that for a system vibrating in one mode, all the particles move in phase and at the same frequency. They reach their maximum points of displacment at the same instant, and they pass through their equilibrium positions at the same instant. If we were able to take a "freeze-frame" view of carbon dioxide vibrating purely in one of its normal modes, then the directions the atoms are moving at the instant the particles are at their equilibrium positions would serve to describe the nature of the vibration. We could represent these directions of motion by arrows at each atom, and this is a very common way of representing molecular normal modes.

One of the normal modes of carbon dioxide is a bend along the O-C-O axis. The arrow representation is

$$\begin{matrix} \uparrow & & \downarrow & & \uparrow \\ \mathrm{O} & - & \mathrm{C} & - & \mathrm{O} \end{matrix}$$

This represents a vibration where the instantaneous direction of motion at the equilibrium point has the oxygen atoms moving up and the carbon atom moving down. Another normal mode is a *breathing motion* or a *symmetric stretch*:

$$\begin{matrix} \leftarrow & & & & \rightarrow \\ \mathrm{O} & - & \mathrm{C} & - & \mathrm{O} \end{matrix}$$

In this motion, the oxygen atoms move away from the carbon atom in phase. If we may use a classical picture for the moment, we would expect the oxygen atoms to continue to their turning points and then reverse directions and eventually pass through the equilibrium positions again. At that instant, their motions might be represented by

$$\begin{matrix} \rightarrow & & & & \leftarrow \\ \mathrm{O} & - & \mathrm{C} & - & \mathrm{O} \end{matrix}$$

This arrow diagram looks different than the previous one, though it is for the same normal mode of vibration. The freeze-frame picture has been taken at a different instant when the particles are at their equilibrium. Either arrow diagram is a correct representation of the symmetric stretching vibration, and it is important not to consider the two drawings to mean different modes.

The mathematics behind the transformation to normal coordinates, which are represented by the arrow diagrams, provides certain rules that may be used to guess the qualitative form of the normal modes of molecules. First, a normal mode is a vibrational motion, and so any motion along the normal coordinate (following the arrows in the diagrams) must not lead to a rotation or translation of the molecule. For carbon dioxide, the following two diagrams are examples of pure translation and pure rotation.

$$\overset{\rightarrow}{O} - \overset{\rightarrow}{C} - \overset{\rightarrow}{O} \qquad\qquad \overset{\uparrow}{O} - C - \overset{\downarrow}{O}$$

If an arrow picture resembles these to any extent, which means that the motion is partly a rotation or translation, then it is not that of a normal mode.

Normal coordinates are independent coordinates. This means that they are orthogonal just as unit vectors along the x-axis, along the y-axis, and along the z-axis of Cartesian space are orthogonal. The normal coordinates are more complicated and abstract. Even so, we may test for orthogonality in about the same way we determine that the x,y,z unit vectors are orthogonal: The dot product of any two different unit vectors is zero. The dot product of a normal coordinate and another normal coordinate is found by adding up the dot products of the corresponding arrows on each atom. As an example, let us use the other stretching mode of carbon dioxide, called the asymmetric stretch. Its arrow diagram is

$$\overset{\rightarrow}{O} - \overset{\leftarrow}{C} - \overset{\rightarrow}{O}$$

This is orthogonal to the symmetric stretch, as seen by taking the dot product of corresponding arrows on the atoms:

$$\overset{\rightarrow 1}{O} - \overset{\leftarrow 2}{C} - \overset{\rightarrow 3}{O}$$

$$\overset{\rightarrow 4}{O} - C - \overset{\leftarrow 5}{O}$$

Taking each of the arrows to be of unit length, the dot product of arrows 1 and 4 is +1. The dot product of the arrows on the right most atom (3 and 5) is –1. For the carbon atom arrows, the dot product is zero, since carbon is not displaced at all in the course of the symmetric stretch. The sum of the three numbers, 1, 0, and –1, is zero. This is the mathematical statement that these two motions, or these two normal coordinates, are distinct or orthogonal.

For all molecules there will be a mode that is best thought of as a breathing mode. All the atoms will move away from the equilibrium. The symmetric stretch mode of carbon dioxide is its breathing mode. To work out arrow diagrams for all the normal modes of a molecule, we may start with a diagram for the breathing mode. Then, we need to identify all other arrow diagrams that are orthogonal, making sure that the diagrams do not include any rotation or translation. If the molecule has symmetry, we can expect the arrow diagrams to reflect that in some way, and that may help systematize our search for the normal coordinate diagrams. Though this process is not necessarily unique, the qualitative information about the nature and types of vibrations will be correct.

As another example, let us find arrow diagrams for the normal modes of the water molecule. The breathing motion must be a simultaneous stretch of the two O-H bonds. If we represent this with an arrow diagram showing the hydrogens moving and the oxygen staying fixed, we will have a representation of a motion where the center of the mass of the molecule moves. Translation will be part of the motion. So, an arrow must be placed on oxygen to make sure that the motion does not include any translation. The representation of the symmetric stretching vibration is

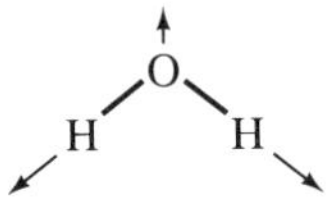

O
H H

The arrow on the oxygen is smaller than the arrows on the hydrogens because it is so much heavier; the center of mass will remain in place for a displacement of oxygen that is relatively smaller than the hydrogen displacements. We may expect a bending motion, and again, an arrow must be used for oxygen to ensure that the motion does not include any translation or rotation.

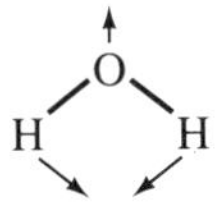

O
H H

An orthogonality test will show that this is an acceptable mode because it is orthogonal to the symmetric stretch. The number of modes is 3N − 6 or 3. So, the remaining mode will be represented by a set of arrows that describe a motion that is orthogonal to both the symmetric stretch and the bend. After some thought, we may realize that this mode is

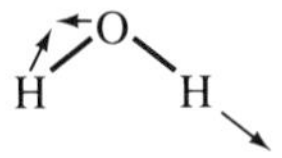

O
H H

This is an asymmetric stretch.

The normal mode pictures are useful in several ways. One way is to roughly anticipate the relative vibrational frequencies. If we consider a classical analog of carbon dioxide, three balls connected by two springs, then the symmetric stretch is stiffer than the asymmetric stretch. The symmetric stretch requires stretching two springs. The asymmetric stretch is like sliding the center ball back and forth between the two balls on the ends. So, we expect the symmetric stretch to be at a higher frequency than the asymmetric stretch, and usually this is the case.

## 5.7 POLYATOMIC VIBRATIONAL SPECTROSCOPY

The selection rule for a diatomic molecule is that the vibrational quantum number changes by one, at least under the harmonic approximation of the potential. Also, the dipole moment has to change in the course of the vibration or else the transition is forbidden. Carbon monoxide, for instance, has an allowed fundamental transition, whereas $N_2$ does not. The separation of variables that is accomplished with the normal mode analysis says that each mode may be regarded as an independent one-dimensional oscillator. Thus, we may borrow the results for the simple harmonic oscillator to conclude that a transition will be allowed if the vibrational quantum number for a mode changes by one and if the vibrational motion gives rise to a changing dipole moment.

The second condition for spectroscopic transitions among polyatomic vibrational levels requires that we determine if molecule's dipole moment will change in the course of a particular normal mode vibration. Returning to the example of carbon dioxide, we could assign a partial charge to the oxygen atoms – call it $\delta$ – and a partial charge to the carbon, which must be $-2\delta$ for the molecule to be neutral. If the atoms are displaced in the direction of the symmetric stretch mode, the contribution to the dipole moment from one oxygen will cancel the contribution from the other because one will move in the +z-direction and the other in the –z-direction. Thus, in the course of the symmetric stretch, the dipole moment of carbon dioxide is unchanging. This means that transitions where the symmetric stretch quantum number is changed will not be (easily) seen in an infrared absorption spectrum. The asymmetric stretch will have allowed transitions since the dipole moment changes in the course of this vibration. The two oxygens move in the same direction, and so their contributions will not cancel. The bending motion will also give rise to a changing dipole moment.

The diatomic selection rule was found to be $|\Delta n| = 1$ when the potential is strictly harmonic. Other transitions become allowed with an anharmonic potential, but then the $|\Delta n| = 1$ transitions stand out as being stronger. For polyatomic molecules, the selection rule under the assumption of a strictly harmonic potential is $|\Delta n_i| = 1$, while $|\Delta n_j| = 0$ for all the j modes other than the i mode. This means that only one vibrational quantum number can change in any single transition event. Of course, molecules do not have *strictly* harmonic potentials, and just as for diatomics, this selection rule is really telling which transitions will be the strong ones. Other transitions become allowed because of anharmonicity. The transitions that obey this harmonic selection rule and that originate from the ground vibrational state are called *fundamental transitions*, which is the same term used with diatomics. Transitions with $|\Delta n_i| > 1$ that originate in the ground state are called *overtone transitions*. If two or more vibrational quantum numbers change in a transition, it is called a *combination transition*.

The vibrational states of molecules are often designated by a list of the vibrational quantum numbers. The ordering of the list is according to the vibrational frequency. The quantum number for the highest vibrational frequency mode is first in the list. For carbon dioxide, a state would be represented as $(n_1, n_2, n_3)$, where $n_1$ refers to the symmetric stretch, $n_2$ refers to the bend, and $n_3$ refers to the asymmetric stretch. Only three quantum numbers are used for this type of designation because the in-plane bend and the out-of-plane bend are two modes that are degenerate; they have the same frequency, and only one quantum number is used. The ground state is (0,0,0). The following is a representative list of possible transitions and their type, if appropriate.

| | | | |
|---|---|---|---|
| (0,0,0) | $\rightarrow$ | (0,0,1) | fundamental |
| (0,0,0) | $\rightarrow$ | (1,0,0) | fundamental |
| (0,0,0) | $\rightarrow$ | (0,2,0) | overtone |
| (1,0,0) | $\rightarrow$ | (2,0,0) | |
| (0,0,0) | $\rightarrow$ | (1,0,1) | combination |

With the many degrees of freedom in a polyatomic molecule, the many different normal mode frequencies, and the many types of transitions that might be seen, it is evident that polyatomic vibrational spectra may be quite congested and challenging to analyze.

The analysis of a full infrared spectrum of a molecule in the gas phase usually starts by finding the fundamental transitions. Most often, these are the strongest transitions. Next, we may look for *progressions*, which are a set of transitions originating from the same initial state and involving the excitation of one mode by successive quantum steps. For instance, the progression built on the fundamental transition (0,0,0,0) $\rightarrow$ (0,1,0,0) of some hypothetical molecule would be the set of transitions from the ground state to (0,2,0,0), and (0,3,0,0), and so on. It is useful to look for these overtone transitions because, to the extent that the harmonic picture holds, we expect them to be found in the spectrum at $2\omega_i$, $3\omega_i$, and so on, where $\omega_i$ is the fundamental transition frequency. Generally, the strongest fundamental transitions will have the strongest overtone progressions. Combination bands, or the low-resolution peaks in the spectrum from combination transitions, may be identified by matching the frequencies with sums and differences of fundamental transition frequencies, and of course, allowing for the fact that the harmonic picture will not hold perfectly for real molecules.

Another type of infrared absorption band is called a *hot band*, and it arises from any transition that does not originate in the ground vibrational state. It is known that the population of excited states will grow as a sample is warmed. In a room temperature sample, it is often possible for the population of a low-lying excited vibrational state to be large enough for transitions to originate from this state and to be detectable. Hot band transitions may occur throughout the spectrum. For instance, a hot band corresponding to the transition (1,0,0) $\rightarrow$ (2,0,0) for a linear triatomic molecule would likely be at a frequency very close to that of the fundamental transition (0,0,0) $\rightarrow$ (1,0,0). The transition moments would be almost the same for the two transitions, and so it is only their populations that lead to any difference in the spectrum. That fact can be exploited: If the temperature of the sample is lowered and the spectrum taken again, the hot band's intensity will diminish relative to a band that originates in the ground vibrational state. This is the consequence of diminishing the excited state's population with decreasing temperature. It is a common

TABLE 5.2 Characteristic Vibrational Stretching Frequencies of Certain Functional Groups. [a]

| | | | |
|---|---|---|---|
| C=O | 1700 cm$^{-1}$ | C≡N | 2200 cm$^{-1}$ |
| C-O | 1100 cm$^{-1}$ | C-N | 1200 cm$^{-1}$ |
| C=C | 1650 cm$^{-1}$ | N-H | 3400 cm$^{-1}$ |
| C≡C | 2200 cm$^{-1}$ | O-H | 3500 cm$^{-1}$ |
| C-H | 3000 cm$^{-1}$ | | |

[a] The characteristic frequencies are ranges around these values, and the ranges sometimes amount to a few hundred wavenumbers.

practice, in fact, to remove hot band congestion from a spectrum by cooling the sample.

There are vibrational transition frequencies which turn out to be characteristic of specific types of chemical bonds. For instance, the chemical bond between carbon and oxygen in a carbonyl functional group is known to be only slightly affected by what the carbonyl is attached to. The C–O bonding in formaldehyde is not sharply different than that in acetone. This means that the force constant for stretching the carbon-oxygen bond will be similar in both molecules. In itself, this does not imply that there will be similar vibrational frequencies for the two molecules. Even within the harmonic picture, we expect the vibration of the carbonyl to be coupled to the rest of the molecule, and that differs from one species to another. Even so, there are similar vibrational frequencies. The frequencies that are associated with functional groups are called *characteristic frequencies*. These are the frequencies at which to expect vibrational transitions in molecules that contain the given functional group. Table 5.2 lists some characteristic frequencies. There is a range for each characteristic frequency because of the effect of coupling with the rest of the molecule. Characteristic frequencies are a useful tool in chemical analysis. A sample of an unknown that exhibits a strong absorption at 1700 cm$^{-1}$, for instance, quite probably contains a carbonyl group in its structure since that is within the range of the characteristic frequency of carbonyl stretching.

Finally, there is a fascinating vibrational motion in certain molecular systems called *inversion* or *interconversion*. An example of inversion is exhibited by the ammonia molecule. It is a pyramid-shaped molecule at its

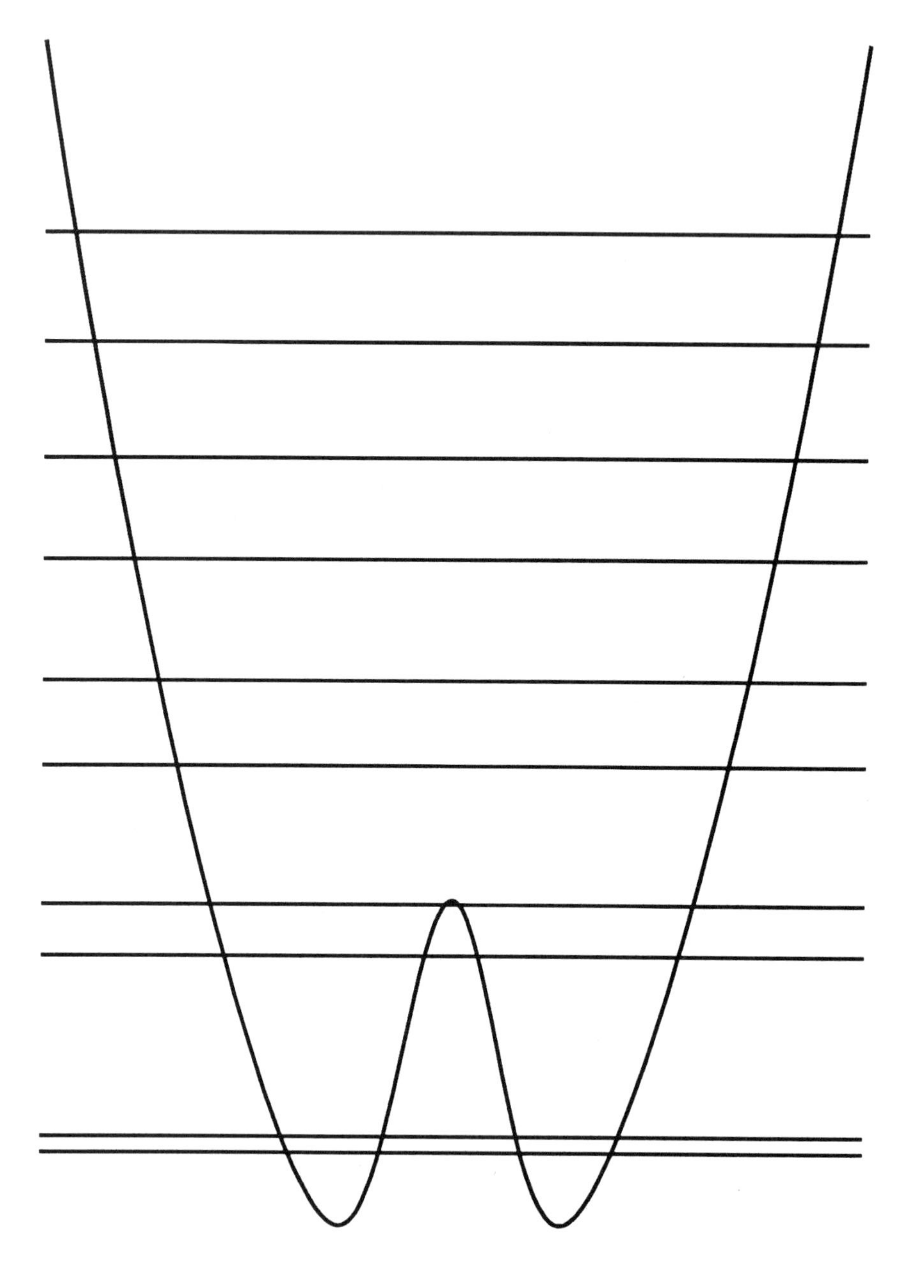

**FIGURE 5.11** The ten lowest-lying vibrational energy levels of a one-dimensional system experiencing a double-minimum potential. The potential is harmonic away from the middle, and the highest levels shown have an energy spacing characteristic of a harmonic oscillator. However, the energies of the first six states are significantly perturbed by the bump in the potential. The perturbation brings the energies of pairs of states closer together.

equilibrium. Let us imagine it as three protons in a plane to the left of the nitrogen, so that an umbrella-opening type of vibration moves the plane of the protons to the right and the nitrogen to the left. If this is continued, a point is reached where the protons and the nitrogen are all in the same plane. This is called the inversion point. If the motion continues still further, the protons will be in a plane to the right of the nitrogen. But then, the molecule will have the same shape as it did originally. It will have achieved a structure that is equivalent to the original structure in terms of the internal structural parameters. The potential energy will have a unique form along this inversion pathway. It will be at a maximum at the inversion point and will have two equivalent minima corresponding to the ammonia pyramid pointing left and pointing right. This is a double-well potential and an example was shown in Fig. 4.2.

The energy levels of a one-dimensional double-well problem are interesting, though working them out can be involved. As mentioned in Chap. 4, perturbation theory may be useful if the inversion barrier is low in energy relative to the ground vibrational state. And of course, a one-dimensional analysis means that motion along the inversion pathway has been separated from other vibrational motions, and that is an approximation. Fig. 5.11 presents calculated energy levels for a particular double-well potential. The effect of the barrier is seen by contrasting the energy level spacing with that of a harmonic oscillator. Instead of evenly spaced levels, the perturbed system has levels brought closer together in pairs. The higher levels show the least effect. The lowest pairs of levels may be brought extremely close by a suitable barrier, and often the separation energy between such levels is termed a *splitting*.

## Exercises

1. Apply perturbation theory in a basis of harmonic oscillator functions to arrive at Eq. (5-23).
2. Evaluate the integral in Eq. (5-27) for the specific case of $J = 2$, $J' = 3$, and $M = M' = 2$.

3. Verify Eq. (5-27) by writing the product, cos θ $Y_{J'M'}$, as a linear combination of other spherical harmonic functions and then using the orthogonality of the spherical harmonics to evaluate the integral.
4. The equilibrium bond length of LiH is 1.595 Å. Find the equilibrium rotational constant in $cm^{-1}$ for LiH and the isotopic forms, LiD and $^{6}LiH$.
5. Assume that it is possible to measure a vibrational transition frequency for a diatomic for which the J quantum number remains unchanged at zero. And assume that there is some diatomic for which the following transition frequencies (in $cm^{-1}$) are then obtained: 1600 for n = 0 to n = 1, 3100 for n = 0 to n = 2, and 4450 for n = 0 to n = 3. Find values for the vibrational frequency, the anharmonicity constant, and the second anharmonicity constant on the basis of these frequencies. Next, compare these values with those obtained for the vibrational frequency and the anharmonicity constant if it is assumed that the second anharmonicity constant, $\omega_e y_e$, is zero.
6. Spectroscopic constants of CO are given in Table 5.1. Find the equilibrium bond length. Then, predict the transition frequencies for the first several P- and R-branch lines in the fundamental absorption band of $^{13}C^{17}O$.
7. The fundamental infrared band of HCl is complicated because the natural isotopic abundance of chlorine means that transitions for two isotopes may be readily observed. That is, unless isotopically purified, the spectrum of an HCl sample is actually a superposition of the spectra of two isotopes of HCl. Which are they? For the first several P- and R-branch lines, calculate the separation in $cm^{-1}$ for the corresponding lines in the spectra of the two isotopic forms. Make a sketch of the expected form of the fundamental band spectrum.
8. Using data in Table 5.1, calculate the band center frequency (i.e., the average of the frequency of the first P-branch and first R-branch lines) for the n = 0→1, n = 1→2, n = 2→3 and n = 3→4 vibrational bands of LiH.
9. Repeat the derivation of Eqs. (5-30) and (5-31) with the inclusion of an energy term, $D\,[\,J\,(\,J+1\,)\,]^2$, associated with centrifugal distortion.
10. From the following general energy level expression for a diatomic,

$$\tilde{E}_{n,J} = (n+\frac{1}{2})\,\tilde{\omega} + \tilde{B}\,J\,(J+1) - \tilde{\alpha}\,(n+\frac{1}{2})\,J\,(J+1) - \tilde{D}\left[J\,(J+1)\right]^2$$

find the branch separation for the fundamental vibrational band in terms of $\tilde{\omega}$, $\tilde{B}$, $\tilde{\alpha}$ and $\tilde{D}$. That is, develop an expression for the difference in transition energies of the two transition lines closest to the band center.

*11.* Assume that the vibrational spectrum of LiH is well represented by the following energy level expression.

$$\tilde{E}_{n,J} = 1405.65\,(n+\frac{1}{2}) - 23.20\,(n+\frac{1}{2})^2 + 7.513\,J\,(J+1)$$

$$- 0.213\,(n+\frac{1}{2})\,J\,(J+1) - 0.01\left[J\,(J+1)\right]^2$$

Find the equilibrium bond length. Next, find the corresponding energy level expression for LiD, taking the potential energy function to be the same as for LiH.

*12.* Using standard bond length values as an estimate of the true bond lengths in HCCH and FCCH, obtain the rotational constants of these two molecules in MHz.

*13.* Shown here are sketches of the normal modes of acetylene. Which modes are infrared-active?

←←→→ HCCH　　←→←→ HCCH　　←→→← HCCH

↑ ↑ HCCH ↓↓　　↑ ↑ HCCH ↓ ↓　　(in plane and out of plane)

*14.* Should be a qualitative difference in the normal modes of vibration, both in the form of the modes and in their infrared activity, of NNO and of $CO_2$.

*15.* Make a sketch using arrows to show a reasonable guess of the qualitative form of the normal modes of vibration for

*a.* formaldehyde　　*b.* ethylene　　*c.* hydrogen peroxide

*16.* To give an idea of what is typical for small molecules, estimate the zero-point energies of the water molecule, of acetylene, and of formaldehyde either on the basis of actual vibrational frequencies or on the basis of characteristic infrared frequencies.

17. A characteristic H-C stretching frequency is 3000 $cm^{-1}$. Let us assume that this is the frequency for the normal mode of HCN that looks most like a stretch of the H-C bond. This mode might be modelled as a pseudo diatomic vibration if we were to think of the CN as one "atom" and the H as the other. Within this model, we may estimate the effects of certain changes to the molecule as if they were simply changes to the mass of the pseudo diatomic. Doing that, what would be the frequency for this vibration if $^{13}C$ were substituted for $^{12}C$? What would be the frequency if the group C-CN were substituted for the N? (This second case offers an idea of why, and to what extent, frequencies are "characteristic" of the bonding environment.)
18. Consider a double well potential of the form $V(x) = x^2/2 + 3/2\, e^{-(\beta x)^2/2}$. If a particle of mass m = 1 were to experience this potential, what would be the energies of the first five states from the following treatments?
    i. Zero-order perturbation energies with the exponential part of the potential being the perturbation
    ii. First-order perturbation theory
    iii. Second-order perturbation theory

## Bibliography

(Also see books listed at the end of Chaps. 3 and 4.)


1. G. W. King, *Spectroscopy and Molecular Structure* (Holt, Rinehart and Winston, New York, 1964). This provides a very comprehensive introduction to vibrational-rotational spectroscopy as well as other types of molecular spectroscopy.
2. G. M. Barrow, *Introduction to Molecular Spectroscopy* (McGraw-Hill, New York, 1962). This is a very useful text at an introductory to intermediate level.
3. W. S. Struve, *Fundamentals of Molecular Spectroscopy* (John Wiley and Sons, New York, 1989). This is an intermediate to advanced level text that will be helpful for further study.

4. J. M. Hollas, *Modern Spectroscopy* (John Wiley and Sons, New York, 1987). This is a valuable text on both the theory and laboratory techniques of spectroscopy.
5. E. B. Wilson, Jr., J. C. Decius, and P. C. Cross, *Molecular Vibrations. The Theory of Infrared and Raman Vibrational Spectra* (Dover Publications, New York, 1980). This is a thorough, advanced level treatment of vibrational spectrosocopy.
6. G. Herzberg, *Molecular Spectra and Molecular Structure. I. Spectra of Diatomic Molecules* (Von Nostrand Reinhold, New York, 1950). This is a research level volume that may be consulted for further study on the spectroscopy of diatomics. It also includes a compilation of spectroscopic data.
7. W. H. Flygare, *Molecular Structure and Dynamics* (Prentice-Hall, Englewood Cliffs, New Jersey, 1978).

# Chapter 6

# Electronic Structure

*Quantum mechanics offers a detailed glimpse of how atoms and molecules exist. It provides a means for understanding how electrons are distributed and how they participate in the formation of chemical bonds. This is the problem of electronic structure, using the Schrödinger equation to find wavefunctions for electrons in atoms and molecules. The atom with the fewest electrons, the hydrogen atom, serves as an important model problem, and the quantum mechanical analysis of the hydrogen atom is carried out in detail in this chapter. From that, we explore the qualitative features of the structure of more complicated atoms and then molecules.*

## 6.1 THE HYDROGEN ATOM

The quantum mechanical description of a hydrogen atom is the starting point for understanding the electronic structure of atoms and of molecules. It is a problem that can be solved analytically, and it is useful to work through the details. The hydrogen atom consists of two particles, an electron and a nucleus, which is a proton. Since the proton mass is about 2000 times that of an electron, the proton would be expected to make small excursions about the center of mass of the atom relative to any excursions of the very light electron. That is, we expect the electron to be moving quickly about, and in effect, orbiting the nucleus.

The Schrödinger equation for this two-body problem starts out the same as the general two-body Schrödinger equation [Eq. (5-1)]; however, the potential function, V(r), will be different from that of the vibrating-rotating diatomic molecule. It is an electrostatic attraction of two point charges and its form is

$$V(r) = -\frac{Ze^2}{r} \tag{6-1}$$

where Z is the integer nuclear charge, which is +1, and e is the size of the fundamental charge of an electron. This interaction is the product of the charges, in this case +Ze for the nucleus and –e for the electron, divided by the distance between them, r.

The potential function for the hydrogen atom is only dependent on the spherical polar coordinate r. Thus, the separation of variables carried out in going from Eq. (5-1) to Eq. (5-5) remains valid for this problem. That means we immediately know that the wavefunctions for the hydrogen atom will consist of some type of radial function, R(r), times a spherical harmonic function, $Y_{lm}(\theta,\phi)$. To find the radial function, we start with Eq. (5-5) and use the potential function of Eq. (6-1).

$$-\frac{\hbar^2}{2\mu}\left(\frac{2}{r}\frac{\partial R}{\partial r} + \frac{\partial^2 R}{\partial r^2} - \frac{l(l+1)}{r^2}R\right) - \frac{Ze^2}{r}R = ER \tag{6-2}$$

$\mu$ is the reduced mass of the two-body system. Notice that with the sizable difference in mass of the two particles, the reduced mass is very nearly equal to the mass of the lighter particle, the electron.

$$\mu = \frac{m_e M_P}{m_e + M_P} = \frac{1836.15\ m_e^2}{m_e + 1836.15\ m_e} = \frac{1836.15\ m_e}{1837.15} \cong m_e$$

$M_P$ is the proton mass and $m_e$ is the electron mass.

In the vibrating-rotating two-body system, it was appropriate to approximate the $l(l+1)$ term in the radial Schrödinger equation; however, that would not be appropriate for the hydrogen atom. The separation distance between the electron and the proton, which is given by the coordinate r, can and does vary widely in the hydrogen atom states, and so there is no basis for using the truncated power series expansion, as was done for the vibrating-rotating diatomic. Fortunately, the differential equation in Eq. (6-2) has known solutions. There are an infinite number of these solutions for each particular value of the quantum number $l$, and we introduce a new quantum number, n, to distinguish these solutions. Both $l$ and n must label the different eigenfunctions, e.g., $R_{nl}(r)$. A condition on this new quantum number, n, that comes about in solving the differential equation is that for some choice of $l$, n may only take on the values $l+1$, $l+2$, $l+3$, and so on. This condition can be inverted so as to relate the value of $l$ to n. Then, we have that for a given choice of n, $l$ can be only 0, 1, 2, ... , or n − 1. Also, n must be a positive integer.

From the separation of variables, we now have that the wavefunctions for the hydrogen atom are

$$\psi_{nlm}(r,\theta,\phi) = R_{nl}(r)\, Y_{lm}(\theta,\phi) \tag{6-3}$$

And the quantum numbers that distinguish the possible states must satisfy these conditions:

$$n = 1, 2, 3, \ldots \tag{6-4a}$$

$$l = 0, 1, 2, \ldots, n-1 \tag{6-4b}$$

$$m = -l, -l+1, \ldots, l-1, l \tag{6-4c}$$

The radial functions, $R_{nl}(r)$, that are the eigenfunctions of Eq. (6-2) may be constructed from a set of polynomials called the *Laguerre polynomials*. The Laguerre polynomial of order k may be generated by

$$L_k(z) = e^z \frac{d^k}{dz^k}(z^k e^{-z}) \tag{6-5}$$

*Associated Laguerre polynomials* may be generated by,

$$L_k^j(z) = \frac{d^j}{dz^j} L_k(z) \tag{6-6}$$

The radial functions, expressed in terms of associated Laguerre polynomials are

$$R_{nl}(r) = \sqrt{\frac{(n-l-1)!}{2n\,[(n+l)!]^3}}\; e^{-\rho/2}\, \rho^l\, L_{n+l}^{2l+1}(\rho) \tag{6-7}$$

where $\rho$ is the variable r scaled by the constants Z, $\mu$, e, and $\hbar$ that are in the Schrödinger equation and by the quantum number, n.

$$\rho \equiv \frac{2 Z \mu e^2}{n \hbar^2}\, r \tag{6-8}$$

The square root factor in Eq. (6-7) is for normalization over the range from $r = 0$ to infinity. Recall that in spherical polar coordinates the volume element is $r^2\, dr\, \sin\theta\, d\theta\, d\phi$, and so the normalization of the radial equations obeys the following:

$$\int_0^\infty R_{nl}^2(r)\, r^2\, dr = 1$$

Table 6.1 lists the explicit forms for several of these radial functions.

When the functions of Eq. (6-7) are used in Eq. (6-2), the energy eigenvalue associated with a particular $R_{nl}(r)$ function is found to be

$$E_{nl} = -\frac{\mu Z^2 e^4}{2\hbar^2 n^2} \tag{6-9}$$

Thus, the energies of the states of the hydrogen atom depend only on the quantum number n. The lowest energy state is with $n = 1$, and for this state, the energy given by Eq. (6-9) is $-109{,}678\ cm^{-1}$. The next energy level occurs with $n = 2$, and this energy is one fourth ($2^{-2}$) of the lowest energy or $-27{,}420\ cm^{-1}$. Next is the energy level for states with $n = 3$. This energy is one ninth ($3^{-2}$) of the lowest energy or $-12{,}186\ cm^{-1}$. At $n = 100$, the energy of the hydrogen atom is $-11\ cm^{-1}$. Clearly, as n approaches infinity, the energy approaches zero. This limiting situation corresponds to the ionization of the atom; the electron is then completely separated from the nucleus, and there is zero energy of interaction.

Since the energy of the hydrogen atom depends only on n, and since, according to Eq. (6-4), there may be several states with the same n, then the states of the hydrogen atom may be degenerate. We can use the rules of Eq. (6-4) to see that the degeneracy of each level is $n^2$, as in Table 6.2.

TABLE 6.1 Hydrogen Atom Radial Functions.

| n, $l$ | | Radial Function $\left[\rho = 2Z\mu e^2 r/(n\hbar^2) \text{ and } N = (Z\mu e^2/\hbar^2)^{3/2}\right]$ |
|---|---|---|
| n = 1 | $l = 0$ | $R_{10}(r) = N\,2e^{-\rho/2}$ |
| n = 2 | $l = 0$ | $R_{20}(r) = \dfrac{N}{2\sqrt{2}}(2-\rho)\,e^{-\rho/2}$ |
| | $l = 1$ | $R_{21}(r) = \dfrac{N}{2\sqrt{6}}\rho\,e^{-\rho/2}$ |
| n = 3 | $l = 0$ | $R_{30}(r) = \dfrac{N}{9\sqrt{3}}(6-6\rho+\rho^2)\,e^{-\rho/2}$ |
| | $l = 1$ | $R_{31}(r) = \dfrac{N}{9\sqrt{6}}(4\rho-\rho^2)\,e^{-\rho/2}$ |
| | $l = 2$ | $R_{32}(r) = \dfrac{N}{9\sqrt{30}}\rho^2\,e^{-\rho/2}$ |
| n = 4 | $l = 0$ | $R_{40}(r) = \dfrac{N}{96}(24-36\rho+12\rho^2-\rho^3)\,e^{-\rho/2}$ |
| | $l = 1$ | $R_{41}(r) = \dfrac{N}{32\sqrt{15}}(20\rho-10\rho^2+\rho^3)\,e^{-\rho/2}$ |
| | $l = 2$ | $R_{42}(r) = \dfrac{N}{96\sqrt{5}}(6\rho^2-\rho^3)\,e^{-\rho/2}$ |
| | $l = 3$ | $R_{43}(r) = \dfrac{N}{96\sqrt{35}}\rho^3\,e^{-\rho/2}$ |
| n = 5 | $l = 0$ | $R_{50}(r) = \dfrac{N}{300\sqrt{5}}(120-240\rho+120\rho^2-20\rho^3+\rho^4)\,e^{-\rho/2}$ |
| | $l = 1$ | $R_{51}(r) = \dfrac{N}{150\sqrt{30}}(120\rho-90\rho^2+18\rho^3-\rho^4)\,e^{-\rho/2}$ |
| | $l = 2$ | $R_{52}(r) = \dfrac{N}{150\sqrt{70}}(42\rho^2-14\rho^3+\rho^4)\,e^{-\rho/2}$ |
| | $l = 3$ | $R_{53}(r) = \dfrac{N}{300\sqrt{70}}(8\rho^3-\rho^4)\,e^{-\rho/2}$ |
| | $l = 4$ | $R_{54}(r) = \dfrac{N}{900\sqrt{70}}\rho^4\,e^{-\rho/2}$ |

TABLE 6.2 Energy Levels of the Hydrogen Atom.

| Level | Energy[a] | Allowed $l$'s | Allowed m's | Degeneracy ($n^2$) |
|---|---|---|---|---|
| n = 1 | $E_{g.s.}$ | 0 | 0 | 1 |
| n = 2 | $E_{g.s.}/4$ | 0 | 0 | |
| | | 1 | 1, 0, –1 | 4 |
| n = 3 | $E_{g.s.}/9$ | 0 | 0 | |
| | | 1 | 1, 0, –1 | |
| | | 2 | 2, 1, 0, –1, –2 | 9 |
| n = 4 | $E_{g.s.}/16$ | 0 | 0 | |
| | | 1 | 1, 0, –1 | |
| | | 2 | 2, 1, 0, –1, –2 | |
| | | 3 | 3, 2, 1, 0, –1, –2, –3 | 16 |

[a] The energy is given in terms of the ground state energy, $E_{g.s.}$, which is –109,678 $cm^{-1}$.

## 6.2 PROPERTIES OF THE RADIAL FUNCTIONS

Examination of the radial functions in Table 6.1 reveals that except for the $l = 0$ functions, all are zero-valued at $r = 0$ ($\rho = 0$). For $r > 0$, each $R_{nl}$ function has $n - l - 1$ points where the function is zero-valued. These points are roots of the polynomials in the $R_{nl}$ functions, and they are simply the points where the radial functions change sign. They are nodes in the wavefunctions. Fig. 6.1 is a plot of several of these functions. Since the quantum mechanical postulates tell us that the square of a wavefunction is the probability density, it is also interesting to notice the forms of $R_{nl}^2$, which are shown in Fig. 6.2.

One of the first features of the hydrogen atom where observations must be reconciled with the quantum mechanical picture is the size of the atom. The quantum mechanical description gives a probability distribution for finding the electron located about the nucleus. In analogy with a classical mechanical system of two particles in an attractive potential, we

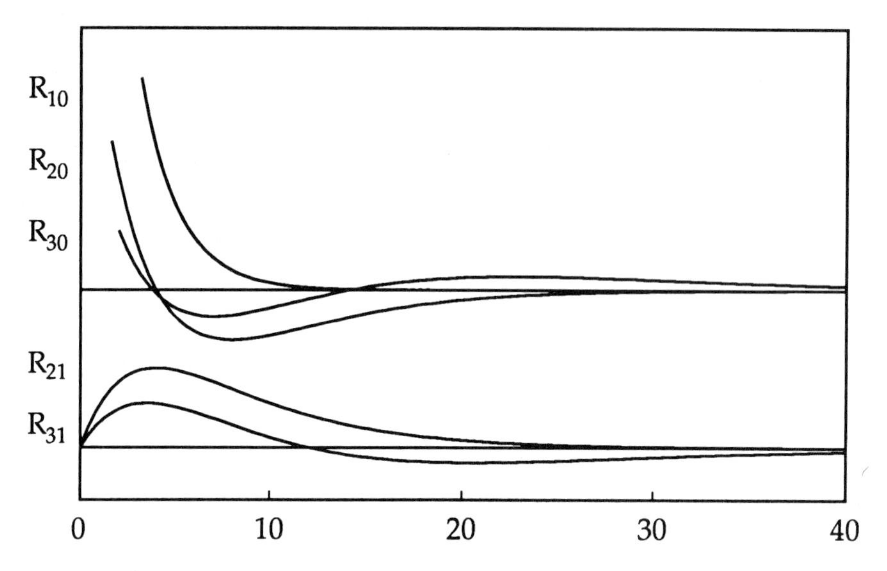


**FIGURE 6.1** Radial functions, $R_{nl}(r)$, of the hydrogen atom. The top three functions are the $l = 0$ functions; the first two $l = 1$ functions are shown below on a different vertical scale. The horizontal scale is in bohr radii, which are 0.52918 Å.

regard this distribution as corresponding to the electron orbiting the nucleus (or both orbiting the center of mass). From Fig. 6.2, we see that the square of the wavefunction dies away smoothly at large r; it does not end abruptly at some particular value of r. This means that there does not exist a finite sphere that entirely encompasses the hydrogen atom and defines its size. Thus, a different notion of atomic size is required, and perhaps the most reasonable one is the average separation distance between the electron and the nucleus. From the quantum mechanical postulates, such an average may be obtained from the expectation value of r. So, for a hydrogen atom state specified by the quantum numbers n, $l$, m,

$$\langle r \rangle_{nlm} = \int \psi^*_{nlm}\, r\, \psi_{nlm}\, r^2\, dr\, \sin\theta\, d\theta\, d\phi$$

$$= \int r^3 R^2_{nl}\, dr \int Y^*_{lm}\, Y_{lm} \sin\theta\, d\theta\, d\phi$$

$$= \beta \frac{n^2}{Z}\left(\frac{3}{2} - \frac{l(l+1)}{2n^2}\right) \qquad (6\text{-}10)$$

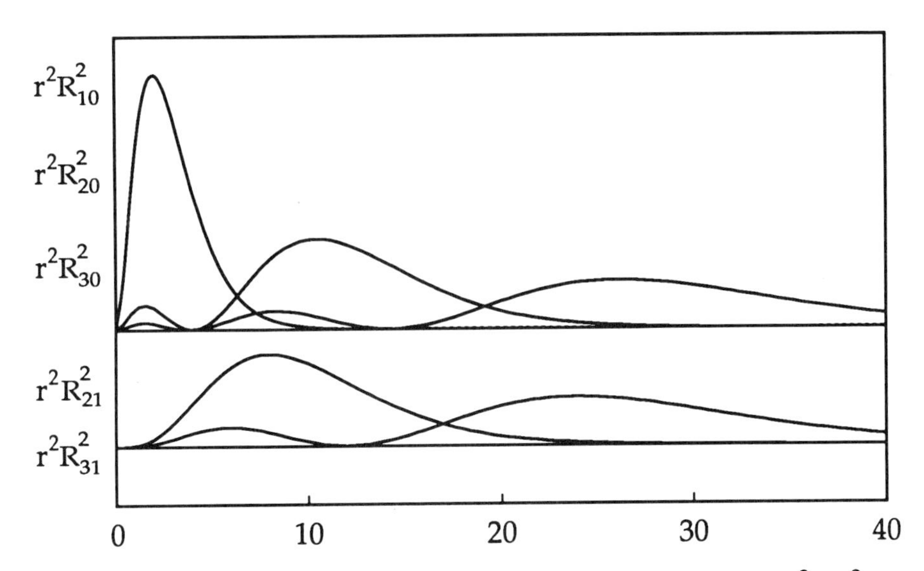


**FIGURE 6.2** Radial probability functions of the hydrogen atom, $r^2R_{nl}^2$. (The factor $r^2$ is included because the radial volume element is $r^2dr$.) The number of nodes increases with both n and *l*.

The integration over the radial coordinate has been carried out by employing special properties of the Laguerre polynomials. The integration over the angular coordinates yields one, since the spherical harmonic functions are normalized. The constant β, called the Bohr radius of the hydrogen atom (also designated $a_0$), is obtained from the reduced mass and two fundamental constants and is equal to 0.529177 Å or $0.529177 \times 10^{-8}$ cm.

$$\beta = \frac{\hbar^2}{\mu e^2} \tag{6-11}$$

For the ground state of the hydrogen atom, the average value of the radial coordinate is $3\beta/2$ or about 0.8 Å. This result fits experience showing that atomic dimensions are on the order of Å. On the other hand, a hydrogen atom in a state very near the ionization limit would be significantly larger. With n = 1000 and $l = 0$, the expectation value for r is about 0.01 cm.

The deviation from the average separation distance can be obtained from the square of the expectation value of r and from the expectation value of $r^2$. The formula that is obtained upon integration of the hydrogenic radial equations with $r^2$ is

$$<r^2>_{nlm} = \frac{\beta^2 n^4}{Z^2}\left(\frac{5}{2} - \frac{3l(l+1) - 1}{2n^2}\right) \tag{6-12}$$

For the ground state, this yields $3\beta^2$, and so the uncertainty in a measurement of r would be $\sqrt{3}\,\beta/2$ or about 0.45 Å.

## 6.3 ORBITAL AND SPIN ANGULAR MOMENTUM

The angular parts of the hydrogen atom wavefunctions are the spherical harmonics, which are, of course, eigenfunctions of the angular momentum operators, $L^2$ and $L_z$. The associated eigenvalues, $l(l+1)\hbar^2$ and $m\hbar$, give the magnitude and the z-component of the angular momentum vector arising from the orbital motion. The orbital motion of an electrically charged particle is a circulation of charge, and that must give rise to the type of magnetic field that is associated with a magnetic dipole source. In classical electromagnetic theory, the magnetic dipole moment,* $\vec{\mu}$, from charge flowing through a circular loop is proportional to the current and to the area of the loop, while its direction is perpendicular to the plane of the loop. We may analyze the hydrogen atom's magnetic dipole by considering it to be a charge, –e, flowing around a loop of some radius r, and then generalizing to the true orbital motion. The area of the loop is $\pi r^2$, and the current is the charge times the frequency at which the charge passes through any particular point on the loop (i.e., the angular velocity, $\omega$, divided by $2\pi$).

$$\mu = (\pi r^2)\left(\frac{-e\omega}{2\pi}\right)\frac{1}{c} = -\frac{e}{2c}r^2\omega \tag{6-13}$$

where c is the speed of light. The angular momentum for a particle moving about a circular loop is the particle's mass times the square of the

* Notice that the Greek letter $\mu$ is used for many things, reduced mass, magnetic dipole moment, and electric dipole moment. While this may be confusing, it should be clear from the context which is meant.

radius of the loop times ω, that is, $mr^2\omega$. By collecting $r^2\omega$ in Eq. (6-13), we can introduce the angular momentum, L, and then generalize to orbital motion by allowing it to be the angular momentum vector of the hydrogen atom:

$$\vec{\mu} = -\frac{e}{2mc}\vec{L} \tag{6-14}$$

The magnetic dipole moment vector is proportional to the orbital angular momentum vector. Since angular momentum will be in the units of ℏ, it is convenient to collect it with the proportionality factor in Eq. (6-14) and define a new constant, $\mu_B$, called the *Bohr magneton.*

$$\mu_B \equiv \frac{e\hbar}{2mc} \tag{6-15}$$

This is the basic unit or measure for electronic magnetic dipole moments in the same sense that ℏ is the measuring unit for angular momentum.

If an external magnetic field is applied to an isolated hydrogen atom, the effect of the field must be incorporated into the quantum mechanical description. This means that the interaction between the magnetic dipole of the orbital motion and the external field must be added to the Hamiltonian. The classical interaction goes as the dot product of the dipole moment and the field. The quantum mechanical operator that corresponds to this classical interaction is easy to determine because of Eq. (6-14). With $\vec{H}$ as the applied magnetic field,

$$\vec{\mu} \bullet \vec{H} = \mu_x H_x + \mu_y H_y + \mu_z H_z$$

$$= -\frac{e}{2mc}\left( L_x H_x + L_y H_y + L_z H_z \right)$$

Letting the orientation of the field define the z-axis in space means that the field components in the x- and y-directions are zero. Thus, the additional term in the Hamiltonian needed to account for an external field is

$$\hat{H}^{int} = -\vec{\mu} \bullet \vec{H} = \mu_B H_z \hat{L}_z / \hbar \tag{6-16}$$

The wavefunctions for the states of a hydrogen atom in an external uniform magnetic field are eigenfunctions of the original Hamiltonian, which we shall now identify as $H^o$, with $H^{int}$ added to it.

We should realize that since the eigenfunctions, $\psi_{nlm}$, of $H^o$ are eigenfunctions of the operator $L_z$, then they must already be eigenfunctions of $H^o + H^{int}$.

$$(\hat{H}^o + \hat{H}^{int})\,\psi_{nlm} = \left(E_n + m\mu_B H_z\right)\psi_{nlm} \tag{6-17}$$

The eigenenergies now have a dependence on the m quantum number. This means that an applied magnetic field will remove the degeneracy of states with the same n and $l$, but with different m quantum numbers. The separation between the levels will increase with the strength of the applied field, according to Eq. (6-17). We can also see from this result why the m quantum number is often referred to as the magnetic quantum number.

The angular form of the orbital functions, $\psi_{nlm}$, or really of the spherical harmonic functions, is interesting. The $l = 0$ or s orbitals are spherically symmetric, and that means they can be represented as spheres. The form of the higher $l$ orbitals is more complicated. Fig. 6.3 shows the form of the m = 0 spherical harmonics. Notice that the number of nodal planes (planes in space where the function is zero) is equal to $l$. Thus, an $l = 1$ m = 0 or $p_0$ function is zero-valued everywhere in the xy-plane. The sign or phase of the function changes from one side of this plane to the other. The $l = 2$ m = 0 or $d_0$ function has two nodal planes.

Certain linear combinations of the spherical harmonic functions are easy to represent and offer a useful picture to keep in mind. The linear combinations and designations for $l = 1$ and $l = 2$ orbitals are

$$Y_{11} + iY_{1\,-1} \rightarrow p_x \qquad Y_{22} + iY_{2\,-2} \rightarrow d_{x^2-y^2}$$

$$Y_{11} - iY_{1\,-1} \rightarrow p_y \qquad Y_{22} - iY_{2\,-2} \rightarrow d_{xy}$$

$$Y_{10} \rightarrow p_z \qquad Y_{21} + iY_{2\,-1} \rightarrow d_{xz}$$

$$Y_{21} - iY_{2\,-1} \rightarrow d_{yz}$$

$$Y_{20} \rightarrow d_{z^2}$$

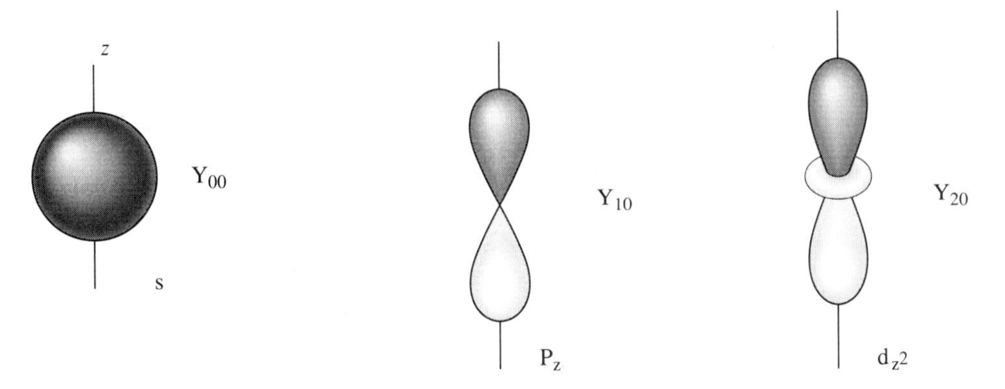


**FIGURE 6.3** Representation of several of the spherical harmonic functions with m = 0. The shapes depict the surface of a contour drawn for some fixed wavefunction amplitude and the alternation in shading designates different phases (different signs) of the wavefunction.

These combinations provide the real parts of the angular functions, and they are represented in Fig. 6.4. Similar combinations can be made for higher $l$ angular momentum functions.

Electrons have an intrinsic angular momentum which gives rise to an intrinsic magnetic dipole moment even if $l = 0$ or even if they are not part of an atom. Very early in the history of quantum mechanics, deflection experiments that measured the magnetic moment of a moving particle by passing it through a magnetic field were able to establish that an electron in an $l = 0$ orbital still possesses a magnetic moment even though $\vec{L} = 0$. Of considerable excitement at the time was the fact that there were two and only two possible z-components of this magnetic moment. Assuming that the magnetic moment was proportional to some angular momentum vector, the questions that arose involved the source of this other type of angular momentum and what was the size of the vector.

The source of the angular momentum is intrinsic to the electron and is referred to as spin, because the spin of a solid body about an axis is a source of angular momentum. However, the term for this intrinsic feature of an electron does not mean we should picture the electron as some mass spinning about its axis. The feature is a more subtle characteristic than that. It can be accounted for with quantum mechanics, but only if the mechanics

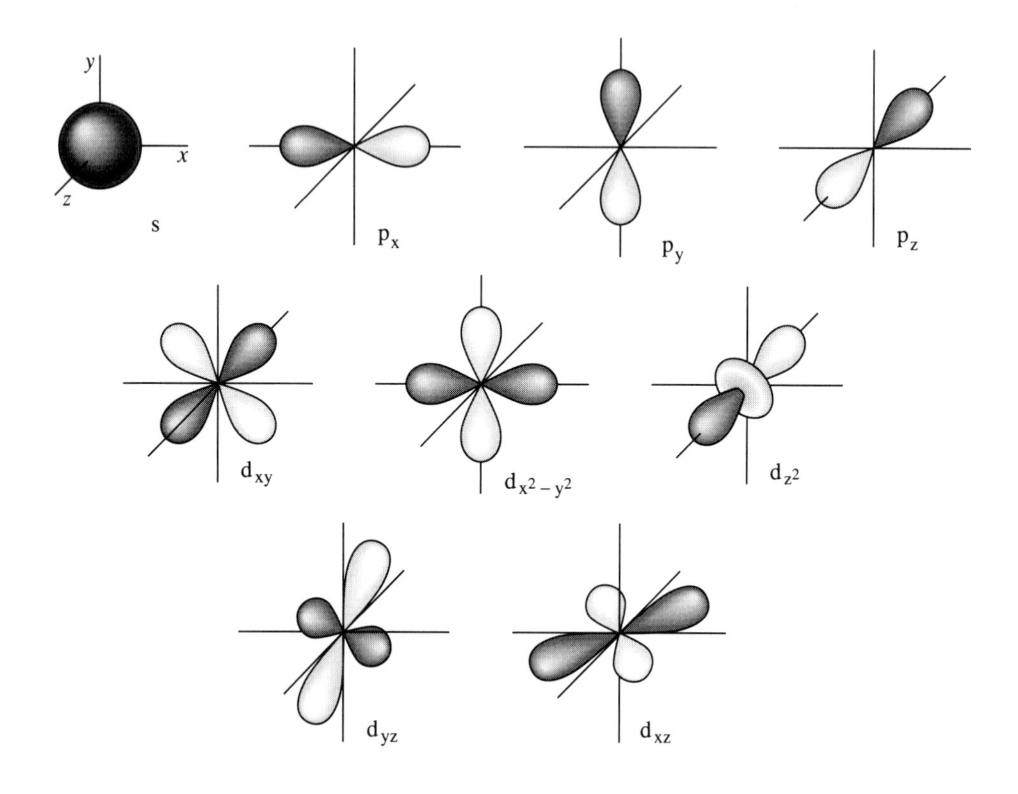


**FIGURE 6.4** Representations of the real parts of the angular functions formed from linear combinations of spherical harmonic functions.

have been adapted for relativistic effects. So, spin is more a name than a description of this property of an electron.

That there are two and only two possible z-axis projections of the electron's intrinsic magnetic dipole, or equivalently, its spin angular momentum vector, is important. For orbital angular momentum, the magnetic quantum number, m, could take on values ranging from $-l$ to $l$. The number of such values, which is the number of different projections on the z-axis, is $2l+1$, an odd number. For there to be an angular momentum that has an even number of projections, the associated quantum number must be a half-integer. Letting s be the quantum number for spin angular momentum, the relation $2s+1=2$ requires that $s=1/2$.

(We say that the spin of an electron is 1/2.) The quantum number that gives the projection of the spin vector on the z-axis is $m_s$, and to avoid confusion at this point, we now designate the orbital angular momentum m quantum number $m_l$.

There are two spin states for an electron. For one, $m_s = 1/2$, and for the other, $m_s = -1/2$. The first state is commonly identified as being the "spin-up" state, and the second is the "spin-down" state. This is because the orientation of the spin vector with respect to the z-axis is in the +z-direction, or up, for the first, and in the –z-direction, or down, for the second. A shorthand designation of these electron spin states is to write $\alpha$ for an electron with $m_s = 1/2$, and $\beta$ for $m_s = -1/2$. $\alpha$ and $\beta$ are meant to be functions in the same way that $\psi$ is a function of r, $\theta$, and $\phi$. However, $\alpha$ and $\beta$ are abstract functions – we will not express them explicitly – of an abstract coordinate, a spin coordinate. $\alpha$ and $\beta$ are orthonormal functions, and that means the following relations hold.

$$\langle \alpha \mid \alpha \rangle = \langle \beta \mid \beta \rangle = 1 \tag{6-18a}$$

$$\langle \alpha \mid \beta \rangle = \langle \beta \mid \alpha \rangle = 0 \tag{6-18b}$$

The integration is over the spin coordinate, and again this is not done as explicit integration.

The magnetic moment that is associated with an electron's spin will interact with an externally applied magnetic field. This magnetic moment is proportional to the spin vector. Analogous to Eq. (6-14) is the following,

$$\vec{\mu} = g_e \frac{e}{2mc} \vec{S} \tag{6-19}$$

This relation differs in an important way from Eq. (6-14). It says that the magnetic dipole moment is proportional to the spin vector with not only a factor of e/2mc but also with an additional factor, $g_e$. (Recall that the proportionality factor for orbital angular momentum is e/2mc.) The additional factor is required because the simple picture of circulation of charge that led to Eq. (6-14) does not apply to the intrinsic spin of an electron. The measured value for the dimensionless constant $g_e$ for a free electron is 2.0023.

## 6.4 ATOMIC ORBITALS AND HYDROGEN ATOM STATES

The wavefunctions, $\psi_{nlm}$, for the hydrogen atom are often referred to as orbitals. Generally, orbitals are any functions of the spatial coordinates of one electron, which in this case means the spherical polar coordinates r, θ, ϕ. The orbitals are named with (1) the value of the principal quantum number, n, (2) a letter associated with the $l$ quantum number, and (3) a numerical subscript, which is the value of $m_l$. The letters* s, p, d, and f were associated with the $l$ quantum number as $s(l = 0)$, $p(l = 1)$, $d(l = 2)$, and $f(l = 3)$. The next is $g(l = 4)$, and from then on the series follows the alphabet. Thus, the ground state orbital of the hydrogen atom is named $1s_0$ or just 1s. The n=2 orbitals are 2s, $2p_1$, $2p_0$, and $2p_{-1}$.

The complete wavefunction of a hydrogen atom is a product of the spatial wavefunction and a spin function, either α or β. So, with the spin of the electron incorporated, the product functions $\psi_{nlm}\alpha$ and $\psi_{nlm}\beta$ (or 1sα, 1sβ, etc.) are referred to as *spin-orbitals*. Spin introduces further degeneracy in the set of eigenfunctions of the original hydrogen atom Hamiltonian, $H^0$. The 1s orbital gives rise to the 1sα and 1sβ spin-orbitals, and so these functions are degenerate functions of $H^0$. Basically, spin doubles the number of states.

The magnetic moments arising from electron orbital motion and from electron spin may interact. This feature of atomic structure, and molecular structure, too, is termed *spin-orbit interaction*. Since the magnetic dipoles due to spin and orbital motion are proportional to their respective angular momentum vectors, the interaction is proportional to the dot product of the angular momentum vectors. This interaction is a small perturbation on the $H^0$ description of the hydrogen atom. We treat it phenomenologically at this point by introducing a proportionality constant, α, rather than by developing a fundamental expression that

* The letters s, p, and d originate in the names sharp, principal, and diffuse. These were the terms that were given to absorption and emission lines in the atomic spectra of alkali atoms on the basis of the appearance of those lines, usually on a photographic plate. Lines of similar type formed series, and it was learned that the transition frequencies measured for a series followed a simple mathematical progression. With quantum mechanics the progressions became understandable consequences of the allowed energy levels and wavefunctions. From this association with types of lines come the orbital letters.

would be used in place of α. (Notice the context distinguishes this use of α from the spin function α.) The perturbing Hamiltonian for spin-orbit interaction, then, is

$$\hat{H}' = \alpha \vec{L} \cdot \vec{S} \tag{6-20}$$

We presume that the spin-orbit interaction constant, α, is to be determined from some measurement.

Spin-orbit interaction implies a coupling of the two "motions" of spin and orbit. From the discussion in Chap. 4, we would expect that this coupling may mix the hydrogenic spin-orbital states with the resulting wavefunctions no longer assured to be eigenfunctions of the $S_z$ and $L_z$ operators. The interaction Hamiltonian of Eq. (6-20) can be rewritten following Eq. (4-48).

$$\hat{H}' = \frac{\alpha}{2}\left[\hat{J}^2 - \hat{L}^2 - \hat{S}^2\right] \tag{6-21}$$

where J refers to the total angular momentum: $\vec{J} = \vec{L} + \vec{S}$. The rules of angular momentum addition give the allowed values of the quantum number J as ranging from $l$ + s downward, in steps of one, to $|l - s|$. For the hydrogen atom's single electron, s = 1/2. Thus, the spin-orbit coupled states of the hydrogen atom will have J quantum numbers equal to $l \pm 1/2$, except if $l$ = 0, then J = 1/2. These spin-orbit coupled states, which we shall designate concisely with the valid quantum numbers in a bra or ket vector, $|nJls\rangle$ or $\langle nJls|$, are eigenfunctions of the operator H' in Eq. (6-21).

$$\hat{H}' \, |nJls\rangle = \frac{\alpha\hbar^2}{2}\left[J(J+1) - l(l+1) - s(s+1)\right] |nJls\rangle \tag{6-22}$$

Thus, the energies of the states after accounting for spin-orbit coupling follow from Eqs. (6-9) and (6-22).

$$E_{nJls} = -\frac{\mu Z^2 e^4}{2\hbar^2 n^2} + \frac{\alpha\hbar^2}{2}\left[J(J+1) - l(l+1) - s(s+1)\right] \tag{6-23}$$

Notice that the energies are subscripted with n, J, $l$, and s since these are the valid quantum numbers for states with spin and orbital motion coupled.

The *spin-orbit splitting* is the energy difference between states that are otherwise degenerate. The case with n = 2 and $l$ = 1 is an example. In the absence of spin-orbit effects, the six hydrogen atom states, $2p_1\alpha$, $2p_1\beta$, $2p_0\alpha$, $2p_0\beta$, $2p_{-1}\alpha$, $2p_{-1}\beta$, are degenerate. These states may be mixed in some way because of spin-orbit interaction, and the resulting states would be distinguished according to the two possible J values, $J = 1 + 1/2 = 3/2$ and $J = 1 - 1/2 = 1/2$. (There are still six states since the $J = 3/2$ coupling is four-fold degenerate and the $J = 1/2$ coupling is doubly degenerate.) Using the formula in Eq. (6-22) we can evaluate the spin-orbit energy for the two possible J values.

$$J = 3/2: \quad E(\text{spin-orbit}) = \frac{\alpha \hbar^2}{2}\left[\frac{3}{2}(\frac{3}{2}+1) - 1(1+1) - \frac{1}{2}(\frac{1}{2}+1)\right]$$

$$= \alpha \hbar^2 / 2$$

$$J = 1/2: \quad E(\text{spin-orbit}) = \frac{\alpha \hbar^2}{2}\left[\frac{1}{2}(\frac{1}{2}+1) - 1(1+1) - \frac{1}{2}(\frac{1}{2}+1)\right]$$

$$= -\alpha \hbar^2$$

$$\text{Energy difference} \quad = -3\,\alpha \hbar^2 / 2$$

The energy difference in these two spin-orbit energies is the splitting. We can see that if the splitting were to be measured spectroscopically, then the value of the spin-orbit interaction constant, $\alpha$, would be known.

When an external magnetic field is applied to the hydrogen atom in a state for which $l > 0$, the quantum mechanical analysis of the energies becomes more complicated. The complete Hamiltonian will include the spin-orbit interaction, the interaction of the orbital magnetic dipole with the field, and the interaction of the spin dipole with the field. Instead of treating this situation generally, consider a special case: If the external field were so strong that the interaction energies with the field were much greater than the spin-orbit interaction, a good approximate description would be to neglect the spin-orbit interaction entirely. (A better approximation would be to include the spin-orbit interaction via low-order perturbation theory.) Physically, the strong field may be thought of as orienting the individual magnetic dipoles and thereby overwhelming their coupling with each other. A very strong field, then, is said to decouple the spin and orbital magnetic dipoles. On the other hand, a very weak field would not decouple the dipoles, but would interact with the net magnetic dipole that results from the sum of the spin and orbital angular momentum vectors. (Again, perturbation theory could be used for a more accurate energetic analysis.) The spectra of atoms and molecules obtained with an applied magnetic field are called *Zeeman spectra.*

## 6.5 ORBITAL PICTURE OF THE ELEMENTS

The Hamiltonians for the Schrödinger equations for many-electron atoms are complicated because of electron-electron interaction. (The like electrical charges repel.) This couples the motions of the different electrons, and that precludes separation of variables. However, if the Schrödinger equation were separable into the coordinates of the different electrons, then the form of the resulting wavefunctions would be products of spin-orbitals. Such a product form can serve as a meaningful approximate description, and it is certainly of qualitative use in understanding the electronic structure of the different elements of the Periodic Table. Within this approximate separation, we may build up the electronic structures of atoms by assigning electrons to specific orbitals or spin-orbitals. These assignments are termed *electron occupancies** in that they indicate the spatial orbitals that are filled by the electrons.

The *Pauli principle* for the electrons of atoms and molecules says that no two electrons can occupy the same spin-orbital, and this is obviously important in building up the orbital picture of the elements. As discussed further on, this principle goes hand in hand with the indistinguishability of the electrons and with their half-integer intrinsic spin. For now, it is important in saying that each distinct spatial orbital (e.g., 1s, 2s, $2p_1$, $2p_0$, $2p_{-1}$) may be occupied by only two electrons. This is because a given spatial orbital can be combined with either an $\alpha$ or $\beta$ spin function to form two and only two different spin-orbitals.

* The term "electron configuration" is also used; however, "electron configuration" can have a more specific meaning, and so the use of "electron occupancy" for the arrangement of electrons in spatial orbitals avoids possible confusion.

A set of spin-orbitals with the same n quantum number is referred to as a *shell*, and the set of spin-orbitals with the same n and $l$ quantum numbers is a *subshell*. The Pauli principle leads to the conclusion that an s type ($l$ = 0) subshell can have an occupancy of at most 2 electrons. A p subshell ($l$ = 1) can have an occupancy of at most 6 electrons, while a d subshell ($l$ = 2) can have an occupancy of at most 10 electrons.

In the hydrogen atom, the energetic ordering of the spatial orbitals is clearly according to the principal quantum number, n. To the extent that this holds for many-electron atoms, we should expect that the orbitals are filled in order of the principal quantum number for the ground states of the elements. That is, as one goes through the Periodic Table, the 1s orbital is expected to be filled first, then the 2s and 2p orbitals, and then the 3s, 3p, and 3d orbitals, and so on.

The interaction between electrons affects the energetically preferred ordering and the shapes of the orbitals in several important ways. (It is appropriate to consider these effects even with the assumed separation of variables that underlies the orbital picture.) The first important consequence of electron-electron interaction is *shielding*. We have already seen in Eq. (6-10) that the average electron-nucleus separation distance for a single-electron atom increases as $n^2$. This simply means that an electron in the 1s orbital is closer on average to the nucleus than an electron in the 2s orbital. The 2s orbital function is closer or tighter than the 3s orbital, and so on. In a lithium atom, where we expect two electrons in the 1s spatial orbital, and one in the 2s orbital, the effective potential that the electron in the 2s orbital experiences is somewhat like a nucleus (Z = +3e) with a tight, negative (–2e) charge cloud surrounding it. In other words, we can view the electron-electron repulsion effects on the 2s electron in this system as if it were a one-electron problem with the positive nucleus being of charge less than +3e, and possibly as little as +e. The nucleus is screened or shielded by the inner two 1s electrons. The immediate consequence is that the lithium atom's 2s orbital is more diffuse, or more spread out radially, than is the 2s electron of the unshielded nucleus, i.e., $Li^{2+}$. It is more diffuse because the effective nuclear charge is smaller.

Shielding, and electron-electron interaction generally, distinguish the energies of electrons in different subshells. For instance, the 2s orbital is usually energetically preferred relative to the 2p orbitals. (The degeneracy in the $m_l$ quantum number remains.) So, the electron occupancy of the carbon atom is $1s^2\ 2s^2\ 2p^2$, which means the 1s and 2s orbitals are fully occupied and there are two electrons occupying 2p orbitals. Were the 2p orbitals preferred, the occupancy would be $1s^2\ 2p^4$, but spectroscopic experiments can unambiguously demonstrate that this is not the occupancy for the ground state of carbon. The ordering of filling orbitals generally follows this pattern throughout the Periodic Table:

$$1s \rightarrow 2s \rightarrow 2p \rightarrow 3s \rightarrow 3p \rightarrow 4s \rightarrow 3d \rightarrow 4p \rightarrow 5s \rightarrow 4d \ldots$$

From this, we can build up the likely electron occupancies of the elements, while remembering that the orbital description implies an approximation involving the separation of variables in the Schrödinger equation.

## 6.6 TERM SYMBOL STATES OF MANY-ELECTRON ATOMS

Transitions between atomic states are seen in absorption and emission spectra, but the spectra of many-electron atoms tend to be complicated because there are numerous states. Angular momentum properties and the splittings associated with spin-orbit interaction can help to interpret atomic spectra. The energetics of the spin-orbit interaction may be understood in the same phenomenological way that was used for the hydrogen atom. That is, we may say that there is an interaction between the magnetic dipoles associated with angular momentum sources and then apply angular momentum coupling rules. The complication, though, is the number of particles and the fact that there may be more than two angular momentum sources.

In light elements, the strongest coupling of magnetic dipoles is between all those associated with orbital motion and between all those associated with spin. In heavy elements, the spin and orbital momenta of individual electrons are the most strongly coupled. So, for light elements, we must apply angular momentum coupling rules to find a total orbital angular momentum vector ($\vec{L}$) and a total electron spin vector ($\vec{S}$). These are then coupled to form a resultant total angular momentum vector ($\vec{J}$). For heavy elements, the orbital and spin vectors of the individual electrons are coupled, just as was done for the single electron of the hydrogen atom, and the resultant vectors from all the electrons are then added to form the total angular momentum vector ($\vec{J}$). We will consider the procedure for the light elements in detail.

For the purpose of working out the angular momentum coupling, electrons in the same subshell are said to be equivalent, while electrons in different subshells are said to be inequivalent. The first situation to consider is that of inequivalent electrons, and the example will be the electron occupancy

$$1s^1\, 2p^1\, 3p^1$$

While this is not an occupancy encountered for the ground states of any of the elements, it could correspond to an excited state of the lithium atom. The task is to add together the orbital angular momenta and to add together the spin angular momenta. Let us use $l_1$, $l_2$, and $l_3$ for the angular momenta of the three electrons. The important rule to apply is that the quantum number for a resultant angular momentum vector may take on values ranging from the sum of the quantum numbers of two sources being combined down to the absolute value of their difference, in steps of one. Since $l_1 = 0$ and $l_2 = 1$, the quantum number for the vector sum ($\vec{L}_{12}$) of these two momenta can only be one.

$$\vec{l}_1 + \vec{l}_2 = \vec{L}_{12} \quad \Rightarrow \quad L_{12} = 1$$

The angular momentum of the third electron is added to $\vec{L}_{12}$. Applying the same rule, we have

$$\vec{l}_3 + \vec{L}_{12} = \vec{L}_{total} \quad \Rightarrow \quad L_{total} = 1+1, \ldots, |1-1| = 2, 1, 0$$

This says that there are three different possibilities for the coupling of the orbital angular momenta, and the three correspond to resultant angular momentum vectors with L quantum numbers of 2 or 1 or 0.

The coupling of the spins is done in the same way: The spin vectors of two electrons are coupled together to yield a resultant vector, and then this is coupled with the spin vector of the next electron. Clearly, the process could be continued for any number of electrons. Furthermore, since the quantum number for electron spin is always 1/2, the number of possibilities is rather limited. Applying the rule for adding angular momenta to the first two spins gives

$$\vec{s}_1 + \vec{s}_2 = \vec{S}_{12} \quad \Rightarrow \quad S_{12} = \frac{1}{2} + \frac{1}{2}, \ldots, \left|\frac{1}{2} - \frac{1}{2}\right| = 1, 0$$

This indicates that the spins of two inequivalent electrons may be coupled in two ways. Adding the third spin gives

$$\vec{s}_3 + \vec{S}_{12} = \vec{S}_{total} \quad \Rightarrow \quad S_{total} = \frac{1}{2} + 1, \ldots, \left|\frac{1}{2} - 1\right| \text{ and } \frac{1}{2} + 0$$

$$= \frac{3}{2}, \frac{1}{2}, \frac{1}{2}$$

Notice that both possibilities for the value of $S_{12}$ are used to find the possible values of $S_{total}$. Also notice that the resulting values of $S_{total}$ include two that are the same. This simply means that there are two distinct ways in which the spin vectors may be coupled that produce a resultant vector with an associated quantum number of 1/2.

The multiplicity associated with a given angular momentum is the number of different possible projections on the z-axis. This is always equal to one greater than twice the angular momentum quantum number; that is, multiplicity(J) $= 2J + 1$. Spin multiplicity is equal to $2S + 1$, and the names singlet, doublet, triplet, quartet, and so on, are attached to states with spin multiplicities of 1, 2, 3, and 4, respectively. From the spin coupling that was just carried out, we can see that two inequivalent electrons may spin-couple to produce either a singlet state (i.e., $S = 0$) or a triplet state (i.e., $S = 1$). Three inequivalent electrons may be coupled to produce a quartet state ($S = 3/2$) or two *different* doublet states ($S = 1/2$).

The magnetic moment associated with the net orbital angular momentum, which we shall now call L instead of $L_{total}$, will interact with the magnetic moment arising from the net spin vector, which shall be S instead of $S_{total}$. With the resultant vector designated J, as before, we have

$$\vec{L} + \vec{S} = \vec{J} \quad \Rightarrow \quad J = L + S, \ldots, |L - S|$$

From a given value for L and for S, a number of J values may result. These are different couplings. We must also realize that there may be several different possibilities, not just one, for the L value and for the S value. All of these are to be included in finding the J values for the resultant states. Continuing with the example of three inequivalent electrons, we may tabulate the possible J's.

| Value of L | Value of S | Resultant J's |
|---|---|---|
| 2 | 3/2 | 7/2, 5/2, 3/2, 1/2 |
| 2 | 1/2 | 5/2, 3/2 |
| 2 | 1/2 | 5/2, 3/2 |
| 1 | 3/2 | 5/2, 3/2, 1/2 |
| 1 | 1/2 | 3/2, 1/2 |
| 1 | 1/2 | 3/2, 1/2 |
| 0 | 3/2 | 3/2 |
| 0 | 1/2 | 1/2 |
| 0 | 1/2 | 1/2 |

It is clear that quite a number of distinct spin-orbit coupled states may be associated with the electron occupancy of this problem.

*Term symbols* are designations used in atomic spectroscopy to designate different electronic states and energy levels. The term symbols encode the values of J, L, and S. For the value of L, a capital letter is written: S for $L=0$, P for $L=1$, D for $L=2$, F for $L=3$, and so on, with G, H, I, etc. The spin multiplicity is written as a presuperscript, and the J value is written as a subscript. The form for these symbols, then, is

$$^{(2S+1)}L_J$$

As an example, if L = 1, S = 1/2, and J = 3/2, the term symbol is $^2P_{3/2}$. In the three electron example we have been using, the table of L, S, and J values is perfect for writing down the term symbols for the states. The first line of the table above would translate into the term symbols $^4D_{7/2}$, $^4D_{5/2}$, $^4D_{3/2}$, $^4D_{1/2}$. The next line would give $^2D_{5/2}$ and $^2D_{3/2}$. These are called *Russell-Saunders* term symbols because Russell-Saunders coupling assumes that the individual orbital angular momenta of the electrons are more strongly coupled than the orbital and spin angular momenta. If the spin-orbit interaction is ignored, the terms symbols are written without the J subscript.

Equivalent electrons, those in the same subshell, are more complicated in the analysis of spin and orbit coupling because of the restrictions of the Pauli principle. Essentially, the different ways in which coupling takes place are restricted. As a simple illustration of this, consider the difference between the electron occupancy $1s^2$ and the occupancy $1s^1\,2s^1$. Because the two electrons in the same 1s orbital must have opposite spins to satsify the Pauli principle, the net spin vector must be zero; that is, S = 0 only. The two inequivalent electrons in the 1s and 2s orbitals, on the other hand, may be spin coupled as a singlet (S = 0) and as a triplet (S = 1) state.

There are a number of schemes available for finding the term symbols for equivalent electron problems. The aim of these schemes is to keep track of the possible electron arrangements among the available spin-orbitals. One easy-to-remember scheme that is also generally applicable starts with making a complete list of all the possible arrangements of electrons in spin-orbitals that correspond to the given electron occupancy. For instance, if there were three electrons in the 2p subshell, they could be arranged among the six spin-orbitals in 20 ways. To work out these 20 arrangements, it is nice to use ↑ and ↓ under the column for a particular spatial orbital to indicate that an electron with $\alpha$ and $\beta$ spin, respectively, occupies that orbital. For three electrons in a 2p subshell, the arrangements that are consistent with the Pauli principle are

| $2p_1$ | $2p_0$ | $2p_{-1}$ |
|---|---|---|
| ↑↓ | ↑ | |
| ↑↓ | ↓ | |
| ↑↓ | | ↑ |
| ↑↓ | | ↓ |
| ↑ | ↑↓ | |
| ↓ | ↑↓ | |
| | ↑↓ | ↑ |
| | ↑↓ | ↓ |
| ↑ | | ↑↓ |
| ↓ | | ↑↓ |
| | ↑ | ↑↓ |
| | ↓ | ↑↓ |

| $2p_1$ | $2p_0$ | $2p_{-1}$ |
|---|---|---|
| ↑ | ↑ | ↑ |
| ↓ | ↑ | ↑ |
| ↑ | ↓ | ↑ |
| ↑ | ↑ | ↓ |
| ↓ | ↓ | ↑ |
| ↓ | ↑ | ↓ |
| ↑ | ↓ | ↓ |
| ↓ | ↓ | ↓ |

Tables of this sort can be set up systematically for any occupancy, including those with both equivalent and inequivalent electrons. The number of arrangements in these tables is equal to the number of states of the system, although each line of the table does not necessarily correspond to a particular state. The number of states is the sum, for all resultant J values, of the multiplicity in J.

Each row in the table above may be regarded as having a specific z-axis projection of the total orbital and total spin angular momentum vectors. In the first line, there are two spin-up electrons and one spin-down. The net $M_S$ quantum number for this line is 1/2, which is the sum of the three individual $m_s$ quantum numbers. The sum of the individual $m_l$ quantum numbers for the three electrons, which is 1 + 1 + 0 = 2 for the first line, is the net $M_L$ quantum number. In other words, the z-axis projections of the individual electron momenta are added.

The information about the possible z-axis projections obtained from the table of spin-orbital arrangements is sufficient to deduce the possible total L and S quantum numbers. To see how this comes about, let us consider just one type of angular momentum. For a given angular momentum quantum number, J, the largest z-axis projection is with M equal to J. If we had a list of M values, but did not know the J value, we would only need to look through the list for the largest M value to know what J is. For instance, from the following list of M values,

3, 2, 2, 1, 1, 0, 0, 0, –1, –1, –2, –2, –3

we would conclude that the biggest possible J value is 3. Then, we would eliminate from this list the M values 3, 2, 1, 0, –1, –2, –3 because they are the M values that go along with J = 3. The remaining list would be

2, 1, 0, 0, –1, –2

The largest of these values is 2, and so there must also be an arrangement that yields J = 2. If we then eliminate the M values of 2, 1, 0, –1, –2 because they are the values that go along with J = 2, the list will be simply

0

This means J = 0. So, we conclude that the original list of M values is uniquely consistent with J values of 3, 2, and 0.

In our table of electron arrangements, the net $M_L$ and $M_S$ quantum numbers will determine the total L and S quantum numbers. The procedure, shown in Table 6.3 for the $2p^3$ example, is to go through the list to find the biggest $M_L$ and $M_S$ and conclude that there is a possible state with corresponding L and S quantum numbers. Then, entries on the list that go along with this (L, S) pair are eliminated, and the process is repeated to find another (L, S) pair. For each (L, S) pair, we write a Russell-Saunders

TABLE 6.3 Development of Term Symbols From the Electron Occupancy $2p^3$. The numbers $M_L$ and $M_S$ are obtained from the sum of the $m_l$ and $m_s$ quantum numbers of the individual electrons in each row. From the complete list of numbers, the pair with the biggest $M_L$ and then the biggest $M_S$ was selected, and this yielded one value for the pair of quantum numbers (L=2, S=1/2). The dots that follow below represent the table entries that are eliminated to account for all the different states with this L and this S. From the remaining entries the biggest $M_L$ and $M_S$ were selected, and the process repeated. The three sets of (L, S) pairs that were obtained dictate the term symbols.

| $2p_1$ | $2p_0$ | $2p_{-1}$ | $M_L$ | $M_S$ | | | | |
|---|---|---|---|---|---|---|---|---|
| ↑↓ | ↑ | | 2 | 1/2 | ⇒ | (L=2, S=1/2) | | |
| ↑↓ | ↓ | | 2 | −1/2 | | • | | |
| ↑↓ | | ↑ | 1 | 1/2 | | • | | |
| ↑↓ | | ↓ | 1 | −1/2 | | • | | |
| ↑ | ↑↓ | | 1 | 1/2 | ⇒ | | (L=1, S=1/2) | |
| ↓ | ↑↓ | | 1 | −1/2 | | | • | |
| | ↑↓ | ↑ | −1 | 1/2 | | • | | |
| | ↑↓ | ↓ | −1 | −1/2 | | • | | |
| ↑ | | ↑↓ | −1 | 1/2 | | | • | |
| ↓ | | ↑↓ | −1 | −1/2 | | | • | |
| | ↑ | ↑↓ | −2 | 1/2 | | • | | |
| | ↓ | ↑↓ | −2 | −1/2 | | • | | |
| ↑ | ↑ | ↑ | 0 | 3/2 | ⇒ | | | (L=0, S=3/2) |
| ↓ | ↑ | ↑ | 0 | 1/2 | | | | • |
| ↑ | ↓ | ↑ | 0 | 1/2 | | | • | |
| ↑ | ↑ | ↓ | 0 | 1/2 | | • | | |
| ↓ | ↓ | ↑ | 0 | −1/2 | | • | | |
| ↓ | ↑ | ↓ | 0 | −1/2 | | | • | |
| ↑ | ↓ | ↓ | 0 | −1/2 | | | | • |
| ↓ | ↓ | ↓ | 0 | −3/2 | | | | • |

term symbol to achieve a concise statement of the different electronic states of the system. From Table 6.3, the first (L, S) pair was L = 2 and S = 1/2. The term symbol to be associated with this result is $^2D$, ignoring the J subscripts, which may be worked out from the values of L and S. The next pair is L = 1 and S = 1/2, and the term symbol is $^2P$. The last pair is L = 0 and S = 3/2, and the term symbol is $^4S$.

With term symbols as the final result of analyzing the spin and orbital angular momenta couplings in atoms, the values of the quantum numbers L, S, and J are established. The spin-orbit interaction energy, obtained with the Hamiltonian of Eq. (6-21) and given in terms of a phenomenological constant $\alpha$, can be obtained in the same way as in Eq. (6-22).

$$E_{JLS}^{\text{spin-orbit}} = \frac{\alpha \hbar^2}{2} \left[ J(J+1) - L(L+1) - S(S+1) \right] \tag{6-24}$$

The constant $\alpha$, though, is not the same from one problem to another.

There is a very useful set of rules, which are named *Hund's rules*, that predict the energetic ordering of the term symbol states that arise from a given electron occupancy. These were first developed empirically on the basis of atomic spectra. The most important rule is that the states will be ordered energetically according to their spin multiplicity, with the greatest spin multiplicity giving the lowest energy. The second rule is that among states with the same spin multiplicity (and arising from the same electron occupancy), the energetic ordering will be according to the L quantum number, the lowest level being that with the greatest L. In the example of the $2p^3$ occupancy used in Table 6.3, Hund's rules would predict that the $^4S$ energy level will be lower than the $^2P$ and $^2D$ levels. Also, the $^2D$ will be lower than the $^2P$. The term symbol for the ground state of the nitrogen atom, which has an occupancy $1s^2\ 2s^2\ 2p^3$, is in fact $^4S$. Additional rules distinguish among the energies according to the J quantum number, but Eq. (6-24) already gives a way of being quantitative about these energy differences.

The spectra of atoms arise from transitions between the possible electronic (term symbol) states. Generally, transitions from the ground state of an atom to an excited state will require the energy of photons in the visible and ultraviolet region of the electromagnetic spectrum. The selection rule is determined from the matrix element of the dipole moment operator and two atomic state wavefunctions. For the hydrogen

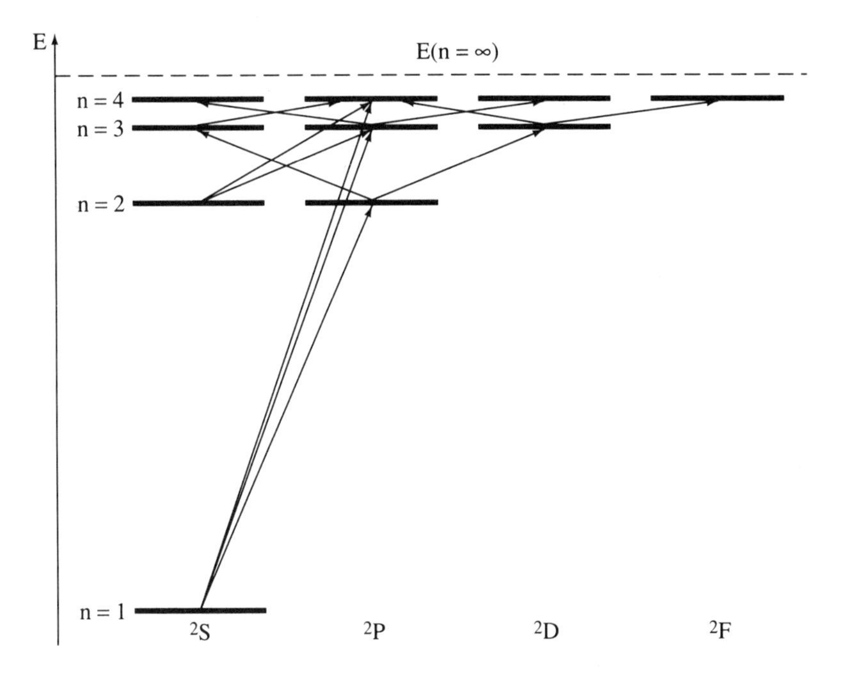


**FIGURE 6.5** The lowest energy levels of the hydrogen atom (horizontal lines) and the allowed transitions. States with the same term symbol have been placed in the same column for clarity.

atom, for which states and allowed transitions are shown in Fig. 6.5, this means that transitions are allowed between a state with quantum numbers n $l$ $m_l$ and n' $l'$ $m_l'$ if the following is nonzero.

$$< \psi_{n\, l\, m_l} \mid \vec{r}\ \psi_{n'\, l'\, m_l'} >$$

This represents three integral values because of the three vector components of $\vec{r}$, and a transition is allowed if any one is nonzero. It turns out that integration over the radial coordinate gives a nonzero result for any choice of n and n', but the angular coordinate integration requires that $l$ and $l'$ differ by 1 for the result to be nonzero. Thus, the selection rule for the hydrogen atom spectrum is stated concisely as

$$\mid \Delta l \mid = 1 \qquad (6\text{-}25)$$

The ground state of the hydrogen atom is $^2S$, and so this selection rule means that transitions would be allowed only to $^2P$ states. From the excited $^2P$ state, transitions to $^2S$ and $^2D$ states would be allowed.

All the states of the hydrogen atom are doublet states. But in many-electron atoms, there may be different spin states. For light elements, where the spin-orbit coupling is weak, the selection rules for atomic spectra are

$$\Delta S = 0 \qquad (6\text{-}26a)$$

$$|\Delta L| = 1 \qquad (6\text{-}26b)$$

Furthermore, the change in L can come about only if the occupancy change corresponds to a change of 1 in the $l$ quantum number of one and only one electron. That is, $\Delta l_i = 1$ while $\Delta l_{j \neq i} = 0$. For example, a change from the occupancy $1s^2\,2s^2\,2p^1$ to the occupancy $1s^2\,2s^1\,2p^2$ is a change in one and only one of the electrons' $l$ values of 1. The term symbol for the first occupancy is $^2P$, and so transitions would be allowed to term symbol states arising from the second occupancy, but only if they were $^2S$ or $^2D$. If we consider the fine structure of the spectra, which means the small energy differences due to spin-orbit interaction, then a selection on J applies:

$$|\Delta J| = 0,\ 1 \qquad (6\text{-}27)$$

This means that J may increase or decrease by one, or it may stay the same.

## 6.7 THE BORN-OPPENHEIMER APPROXIMATION

The electronic structure of molecules is a more complicated problem than the electronic structure of atoms because instead of one central nuclear potential there are several positively charged nuclei distributed in space. Analyzing molecular electronic structure usually begins with an approximate separation of the electronic problem from the problem of nuclear motion. This separation was implicit in the treatment of vibration and rotation in Chap. 5.

The general molecular Schrödinger equation, apart from electron spin effects, is

$$(T_n + T_e + V_{nn} + V_{ee} + V_{en})\Psi = E\Psi \tag{6-28}$$

where the operators in the Hamiltonian are the kinetic energy operators of the nuclei and the electrons, and then the potential energy operators between the nuclei, between the electrons, and between the nuclei and electrons. The explicit forms of these operators are

$$T_n = -\sum_{\alpha}^{\text{nuclei}} \frac{\hbar^2}{2M_\alpha} \nabla_\alpha^2 \tag{6-29}$$

$$T_e = -\sum_{i}^{\text{electrons}} \frac{\hbar^2}{2m_e} \nabla_i^2 \tag{6-30}$$

$$V_{nn} = \sum_{\alpha>\beta} \frac{Z_\alpha Z_\beta e^2}{R_{\alpha\beta}} \tag{6-31}$$

$$V_{ee} = \sum_{i>j} \frac{e^2}{r_{ij}} \tag{6-32}$$

$$V_{en} = -\sum_{\alpha}\sum_{i} \frac{Z_\alpha e^2}{r_{i\alpha}} \tag{5-33}$$

There is a repulsive interaction among the nuclear charges and a repulsive charge-charge interaction among the electrons. However, the interaction potential between electrons and nuclei is attractive since the particles are oppositely charged. This particular interaction couples the motions of the electrons and the motions of the nuclei.

The wavefunctions that satisfy Eq. (6-28) must be functions of both the electron position coordinates and the nuclear position coordinates, and this differential equation is not separable. In principle, true solutions could be found, but the task is surely difficult as this is a formidable differential equation. An alternative is an approximate separation of the differential equation based upon the sharp difference between the mass of an electron and the masses of the nuclei. The difference suggests that the nuclei will be sluggish in their motions relative to the electron motions. Over a brief period of time, the electrons will "see" the nuclei as if fixed in space; the nuclear motions will be relatively slight. The nuclei, on the other hand, will "see" the electrons as something of a blur, given their fast motions.

We may exploit the distinction between sluggish and fast particles by attempting a separation of variables. First, the wavefunction is taken to be (approximately) a product of a function of nuclear coordinates only, $\phi$, and a function of electron coordinates, $\psi$, *at a specific nuclear geometry*. At the specific nuclear geometry, $R=R_o$ (R collectively stands for all nuclear postion coordinates), this is expressed as

$$\Psi(R_o, r_1, r_2, \dots) \cong \phi(R_o)\, \psi^{\{R_o\}}(r_1, r_2, \dots) \qquad (6\text{-}34)$$

Notice the superscript on the electronic wavefunction $\psi$. It designates that this function is for $R = R_o$, a specific point. If this product form of the wavefunction is to be used in the Schrödinger equation, then we need to know something about the effect of the kinetic energy operators. Since $T_e$ acts only on electron coordinates, then $T_e\Psi = \phi T_e\psi$. $T_n$ affects both functions since $\psi$, the electronic wavefunction, retains a dependence on the nuclear position coordinates as parameters. However, the diffuseness of an electronic wavefunction relative to the nuclear positions, again a consequence of mass difference, means that $\phi$ changes more rapidly with R than does $\psi$. In other words, it is reasonable to approximate the effect of $T_n$ on the electronic wavefunction as zero.

$$T_n\, \Psi(R, r_1, r_2, \dots) = T_n\, \phi(R)\, \psi^{\{R\}}(r_1, r_2, \dots) \cong \psi^{\{R\}}(r_1, r_2, \dots)\, T_n\, \phi(R) \qquad (6\text{-}35)$$

With this approximation, the Schrödinger equation is

$$\phi T_e \psi + \psi T_n \phi + \phi\psi(V_{nn} + V_{en} + V_{ee} - E) = 0 \qquad (6\text{-}36)$$

Rearrangement of the terms in this expression leads to

$$\phi\, [\, T_e + V_{ee} + V_{en}\, ]\, \psi + \psi\, [\, T_n + V_{nn} - E\, ]\, \phi = 0 \qquad (6\text{-}37)$$

Of the operators in brackets acting on $\psi$, only one involves the nuclear position coordinates. It is a term giving the electron-nuclear attraction potential. In $\psi$, the nuclear coordinates are treated as parameters, and the same may be done for this operator term since it acts on $\psi$. This allows us to write a purely electronic Schrödinger equation as

$$[\, T_e + V_{ee} + V_{en}^{\{R\}}\,]\, \psi^{\{R\}}(r_1, r_2, \ldots) = E^{\{R\}} \psi^{\{R\}}(r_1, r_2, \ldots) \qquad (6\text{-}38)$$

The energy eigenvalue, E, must also have a parametric dependence on the nuclear positions, as denoted by its superscript. That is, at each different R, there will be a different $V_{en}$, and consequently a different $\psi$ and a different E.

Substitution of Eq. (6-38) into Eq. (6-37) yields

$$\phi E^{\{R\}} \psi^{\{R\}} + \psi^{\{R\}} [\, T_n + V_{nn} - E\,]\, \phi = 0 \qquad (6\text{-}39)$$

Since $\phi$ is a function only of nuclear position coordinates, then the following separation results.

$$[\, T_n + (V_{nn} + E^{\{R\}}) - E\,]\, \phi = 0 \qquad (6\text{-}40)$$

This is the Schrödinger equation for finding $\phi$. The Hamiltonian consists of a kinetic energy operator for the nuclear position coordinates, the repulsion potential between the nuclei, and an effective potential, in the form of $E^{\{R\}}$, that gives the energy of the electronic wavefunction as it depends on the nuclear position coordinates.

The Born-Oppenheimer approximation leads to a separation of the molecular Schrödinger equation into a part for the electronic wavefunction and a part for the nuclear motions, which is the Schrödinger equation for vibration and rotation. The essential element of the approximation is Eq. (6-35) which is that applying the nuclear kinetic energy operator to the electronic wavefunction yields zero. The physical idea is that the light, fast-moving electrons readjust to nuclear displacements instantaneously. This is the reason the approximation produces an electronic Schrödinger equation for each possible geometrical arrangement of the nuclei in the molecule (e.g., each value of R in the diatomic we considered). We need only know the instantaneous positions of the nuclei, not how they are moving, in order to find an electronic wavefunction, at least within this approximation.

The approximation is put into practice by "clamping" the nuclei of a molecule of interest. That means fixing their position coordinates to correspond to some chosen arrangement or structure. Then, the electronic Schrödinger equation [Eq. (6-38)] is solved to give the electronic energy for this clamped structure. After that, perhaps, another structure is selected, and the electronic energy is found by solving Eq. (6-38) once more. Eventually, enough structures might be treated so that the dependence of the electronic energy on the structural parameters is known fairly well. At such a point, the combination of $E^{\{R\}}$ with $V_{nn}$ yields the effective potential for molecular vibration. It is the potential energy for the nuclei in the field created by the electrons.

The separation of the electronic and nuclear motion via the Born-Oppenheimer approximation leads to an important concept, that of the *potential energy surface.* A potential energy surface gives the dependence of the electronic energy, plus nuclear repulsion, on the geometrical coordinates of the atomic centers of a molecule. It is only within the Born-Oppenheimer approximation that we can follow the dependence of the electronic energy on the atomic position coordinates. That only comes about with the separation of the electronic and nuclear problems.

Potential energy surfaces are the conceptual framework for thinking about and analyzing many problems in chemistry. The surface, of course, is the effective potential for vibrations of the molecule. But continued to large atom-atom separations, the surface encodes the energetic information about reactions and bond breaking. It also provides a picture of the pathways for reactions.

A potential energy surface is a representation of a function of all the internal coordinates of a molecular system. For even a simple molecule such as ammonia, the potential is a function of six coordinates, making it difficult to display or to visualize. We may, however, make a graphical representation of a slice through the multidimensional surface. The slice is the potential function with all but one or two of the coordinates fixed at certain chosen values. If all but one are fixed, the slice is a function of only that one coordinate and its representation is a potential curve. If all but two are fixed, the slice is a function of two coordinates.

Functions of two coordinates are easily represented by contour diagrams, and this is one of the generally used ways of presenting potential energy surface slices. An example is shown in Fig. 6.6. The curves that we see in this figure connect or follow equipotential points on the potential energy surface. Cutting across one of these curves means going "uphill" or "downhill" in energy. Usually, we may identify a lowest energy point on a slice through a potential energy surface (i.e., a two-dimensional contour

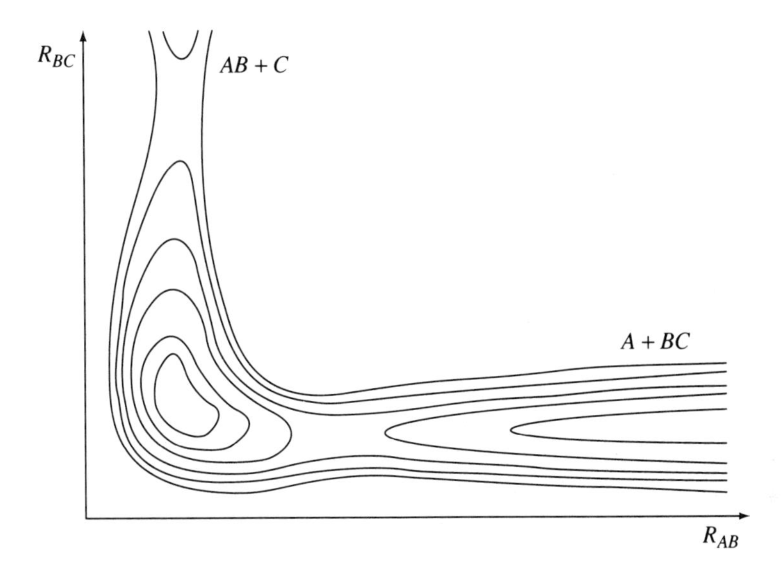


**FIGURE 6.6** A contour diagram of a slice of a hypothetical, but representative, potential energy surface of a triatomic molecule, ABC. The horizontal axis coordinate is the A-B distance, and the vertical axis coordinate is the B-C distance. The bending angle is fixed at 180°. The energy steps between adjacent contour levels are meant to be alike.

plot). The contour plot can only show us that the potential is uphill in two directions away from this point. In the other directions, the ones that were fixed to create the slice, the surface might be sloping upward or downward. Should it be that the surface slopes upward in every direction, then the point is a minimum energy point or *minimum* on the surface. For complicated surfaces, it is possible for there to be several minima, and the one of lowest energy is called the *global minimum*. The global minimum is a point on the surface that corresponds to the *equilibrium structure* of the species.

A slice of a hypothetical potential energy surface of a hypothetical linear triatomic molecule, ABC, is shown in Fig. 6.6. This diagram extends to the regions where the molecule is dissociated into the diatomic AB and atom C and where it is dissociated into the diatomic BC and atom A. The minimum is near the lower left. Stretching either bond, which means increasing either coordinate, is an uphill process. However, there is a unique point in the A-B stretching where the potential will start to be downhill. This point, called a *saddle point*, is where the potential surface is uphill in each direction except one, and in that one direction, the potential is downhill, either forward or backward.

If we follow the contour plot of Fig. 6.6 to the right of the saddle point, we encounter a trough. This is simply a region where the potential is slowly changing, either upward or downward, along one direction. Eventually, this trough will have a flat bottom and walls that do not change with the A-B separation as the point is reached where the A atom is completely removed from the BC diatomic and there is no interaction. At this limit (far right of Fig. 6.6), a cut through this two-dimensional surface for any A-B distance is a potential curve; it is the potential energy as a function of the B-C separation, which is just the stretching potential of the BC diatomic. Likewise, a horizontal cut across the top left of the surface in Fig. 6.6 must be the stretching potential for the AB diatomic.

The bottom of a trough that connects limiting regions such as these with minima or with saddle points is called a *minimum energy path*. That is a good term, especially if we consider the potential surface as if it were a physical surface with real hills and valleys. If we were at the bottom of the trough on the far right of the surface in Fig. 6.6 and wished to "hike" to the equilibrium structure on the lower left, the minimum climb would take us right through the saddle point and then down the valley where the potential minimum is found. That is the minimum energy path. Notice that is does not follow any one coordinate direction. At the outset, it follows the direction of the coordinate of the horizontal axis. Then it twists a bit, and if we follow it further, it will follow the direction of the coordinate of the vertical axis.

*Energetic profiles* for reactions, interconversions, and isomerizations are often given as simple curves of potential energy (vertical axis) versus a *reaction coordinate* (horizontal axis). The reaction coordinate is measured in uniform steps *along* the minimum energy path. In this sense, energy profiles are simply special slices through potential energy surfaces. Again, the Born-Oppenheimer approximation provides the basis for an important chemical concept, and it provides the context for discussing details of reaction energetics.

## 6.8 ANTISYMMETRIZATION OF ELECTRONIC WAVEFUNCTIONS

The electrons moving about in an atom or molecule are indistinguishable particles. The mathematical consequence is that the probability density of an electronic wavefunction of an atom or molecule must be invariant with respect to how the electron coordinates are labelled. With $\vec{r}_i$ being the designation of the position vector of the $i^{th}$ electron, interchange of two electrons' position coordinates means swapping $\vec{r}_i$ and $\vec{r}_j$ in $\psi$. For the probability density to be unchanged, the following must hold for any choice of i and j.

$$\psi(\vec{r}_1, \vec{r}_2, \ldots, \vec{r}_i, \ldots, \vec{r}_j, \ldots)^2 = \psi(\vec{r}_1, \vec{r}_2, \ldots, \vec{r}_j, \ldots, \vec{r}_i, \ldots)^2 \quad (6\text{-}41)$$

For the wavefunction itself, this implies that interchange of a pair of electron coordinates can change the wavefunction only by a factor $e^{i\varphi}$, since $e^{i\varphi}(e^{i\varphi})^* = 1$. This same reasoning applies to a wavefunction of any other kind of indistinguishable particle, and it turns out that the value of $\varphi$ depends on the intrinsic spin of the particles. As will be verified later, for electrons and other half-integer spin particles, $\varphi = \pi$. (For other than half-integer spin particles, such as those with spin = 0, 1, 2, ..., the phase is zero: $\varphi = 0$.) Since $e^{i\pi} = -1$, all electronic wavefunctions must satisfy the following condition.

$$\psi(\vec{r}_1, \vec{r}_2, \ldots, \vec{r}_i, \ldots, \vec{r}_j, \ldots) = -\psi(\vec{r}_1, \vec{r}_2, \ldots, \vec{r}_j, \ldots, \vec{r}_i, \ldots) \quad (6\text{-}42)$$

This is termed an *antisymmetrization* requirement; an electronic wavefunction is antisymmetric (changes sign) with respect to interchange of any pair of electron position coordinates.

To impose Eq. (6-42) on some electronic wavefunction, a special operator, called the antisymmetrizer, can be used. Let us develop the form of this operator, assuming that we have at hand a normalized electronic wavefunction, $\Phi$, for some system, and that $\Phi$ has the form of a product of independent, orthonormal functions, u(r), of the electron coordinates.

$$\Phi(\vec{r}_1, \vec{r}_2, \ldots) = u_1(\vec{r}_1)\, u_2(\vec{r}_2) \ldots$$

The interchange of a pair of position coordinates is a well-defined mathematical operation, as we have in Eqs. (6-41) and (6-42). Thus, we can define an operator that performs that interchange: $P_{ij}$ will be an operator that interchanges the coordinates of some electron, i, with the coordinates of some electron, j. That is,

$$P_{ij}\Phi(\vec{r}_1, \vec{r}_2, \ldots, \vec{r}_i, \ldots, \vec{r}_j, \ldots) = \Phi(\vec{r}_1, \vec{r}_2, \ldots, \vec{r}_j, \ldots, \vec{r}_i, \ldots)$$

From this we may construct a new function, Φ', that is assured to be antisymmetric with respect to interchange of particles 1 and 2:

$$\Phi' = \frac{1}{\sqrt{2}}\left(\Phi - P_{12}\Phi\right) \tag{6-43}$$

Antisymmetrization is tested by applying $P_{12}$ and finding that the negative of the function results:

$$P_{12}\Phi' = \frac{1}{\sqrt{2}}\left(P_{12}\Phi - P_{12}P_{12}\Phi\right) = -\Phi'$$

Notice that $P_{12}$ applied twice means swapping the coordinates and then swapping them back, and that is an identity operation. The square root of 2 in Eq. (6-43) is introduced so that Φ' is normalized if Φ is normalized.

$$<\Phi' \mid \Phi'> = \frac{1}{2} <\Phi - P_{12}\Phi \mid \Phi - P_{12}\Phi>$$

$$= \frac{1}{2}\left[<\Phi \mid \Phi> - <\Phi \mid P_{12}\Phi> - <P_{12}\Phi \mid \Phi> + <P_{12}\Phi \mid P_{12}\Phi>\right]$$

$$= \frac{1}{2}[1 - 0 - 0 + 1] = 1$$

The cross-terms must be zero because of the independence of the two particles' coordinates and the orthogonality of the u(r) functions.

At this point, it is helpful to regard Eq. (6-43) as the application of an operator that antisymmetrizes the function with respect to the interchange of the first two particles.

$$\Phi' = \left\{\frac{1}{\sqrt{2}}\left(1 - P_{12}\right)\right\}\Phi \tag{6-44}$$

This operator is one minus the interchange operator with a factor to ensure normalization.

The next step in imposing Eq. (6-42) on $\Phi$ involves electron 3. The wavefunction must be made antisymmetric with respect to interchanging electron 1 with 3 (i.e., applying $P_{13}$), and to interchanging electron 2 with 3 (i.e., applying $P_{23}$). This can be accomplished by a generalization of the operator in Eq. (6-44) to use $P_{13}$ and $P_{23}$, and then by applying it to $\Phi'$.

$$\Phi'' = \left\{ \frac{1}{\sqrt{3}} \left( 1 - P_{13} - P_{23} \right) \right\} \Phi'$$

$$= \left\{ \frac{1}{\sqrt{3}} \left( 1 - P_{13} - P_{23} \right) \right\} \left\{ \frac{1}{\sqrt{2}} \left( 1 - P_{12} \right) \right\} \Phi$$

We can continue this for all the electrons. For instance, considering the fourth electron will mean using $P_{14}$, $P_{24}$, and $P_{34}$. So, for N electrons, an entirely antisymmetric wavefunction, $\psi$, may be constructed from an arbitrary wavefunction, $\Phi$, by application of a sequence of operators:

$$\psi = \frac{1}{\sqrt{N!}} \left( 1 - P_{1N} - P_{2N} - \ldots P_{N-1,N} \right) \ldots \left( 1 - P_{13} - P_{23} \right) \left( 1 - P_{12} \right) \Phi \quad (6\text{-}45a)$$

$$= A_N \Phi \quad (6\text{-}45b)$$

The sequence of operators and the collective normalization constant may be considered as one overall antisymmetrizing operator, which we will call the *antisymmetrizer* for N electrons, $A_N$.

There are two important properties of the antisymmetrizer. First, if it is applied to a wavefunction that is already properly antisymmetric, it will make no change. The implication of this statement is that if the antisymmetrizer is applied twice to an arbitrary function, the same result will be achieved as if it were applied only once; the second application does not do anything. This is the condition of *idempotency*, and $A_N$ is an idempotent operator. The operator equation that expresses this fact is

$$A_N A_N = A_N \quad (6\text{-}46)$$

(This is by no means a statement that $A_N = 1$.) The second property is that it commutes with the electronic Hamiltonian. It is in the Hamiltonian, of course, that the indistinguishability of the electrons is apparent. An interchange of electron coordinates in the Hamiltonian leaves the Hamiltonian unchanged. That is, $P_{ij}H = H$ for any choice of i and j electrons. With this, it is easy to demonstrate that the antisymmetrizer for a two-electron system commutes with the Hamiltonian:

$$\left[H, \frac{1}{\sqrt{2}}(1-P_{12})\right]\Phi(r_1,r_2) = \frac{1}{\sqrt{2}}\left\{H(1-P_{12})\,\Phi(r_1,r_2) - (1-P_{12})H\Phi(r_1,r_2)\right\}$$

$$= \frac{1}{\sqrt{2}}\left\{H\Phi(r_1,r_2) - H\Phi(r_2,r_1) - H\Phi(r_1,r_2) + H\Phi(r_2,r_1)\right\} = 0$$

It can be proved, as well, that the antisymmetrizer for an N-electron system commutes with the Hamiltonian.

Since the antisymmetrizer commutes with the Hamiltonian, Theorem 4.4 means that wavefunctions can be found that are simultaneously eigenfunctions of the Hamiltonian and of the antisymmetrizer. If the antisymmetrizer is applied to an already-antisymmetric wavefunction, it should give back that wavefunction, and that says merely that the eigenvalue associated with the antisymmetrizer is one. Likewise, if the antisymmetrizer is applied to an arbitrary electronic wavefunction, the resulting function is an eigenfunction of the antisymmetrizer, with eigenvalue of one, because of the idempotency of the antisymmetrizer. Because of this, it is a common approach to use basis functions for variational and perturbative treatments of electronic wavefunctions that are already antisymmetrized.

Spin-orbital functions have been defined for atoms already. For molecules, we may construct one-electron spatial wavefunctions by making linear combinations of atomic orbitals (LCAO) and by other means. Spin orbitals for molecules, then, are products of a spatial function of the spatial coordinates of one electron and a spin function, $\alpha$ or $\beta$, for that electron. The orbital picture of the ground state of the ammonia molecule, for example, has 10 electrons in five spatial orbitals. If these are designated $\phi_1$ through $\phi_5$, then the spin-orbitals are

$$\phi_1\alpha,\ \phi_1\beta,\ \phi_2\alpha,\ \phi_2\beta,\ \phi_3\alpha,\ \phi_3\beta,\ \phi_4\alpha,\ \phi_4\beta,\ \phi_5\alpha,\ \phi_5\beta$$

The orbital picture implies that the wavefunction is a product,

$$\Phi(r_1, \ldots r_{10}) = \phi_1(r_1)\alpha(s_1)\,\phi_1(r_2)\beta(s_2) \ldots \phi_5(r_9)\alpha(s_9)\,\phi_5(r_{10})\beta(s_{10})$$

where the abstract spin coordinates are $s_1$, etc. This, of course, is not antisymmetric, but the antisymmetrizer can be applied to it.

It turns out that the function produced by applying the antisymmetrizer to a product of spin-orbitals can be expressed in the concise and useful form of a determinant (see Appendix I). In explaining this, it is convenient to adopt a special shorthand notation. We will use one symbol for a spin-orbital, and here that symbol will be u. A given spin-orbital is a function of the spin and spatial coordinates of some electron, but instead of writing $u(r_i,s_i)$, we will write u(i); that is, just the electron number will be written since that is sufficient to interpret what is meant. In this shorthand notation, the product function given above for ammonia would be written as

$$\Phi(r_1, \ldots r_{10}) = u_1(1)\, u_2(2)\, u_3(3)\, u_4(4)\, u_5(5)\, u_6(6)\, u_7(7)\, u_8(8)\, u_9(9)\, u_{10}(10)$$

In general, the orbital product form of an N-electron wavefunction is

$$\Phi(r_1, \ldots r_N) = u_1(1)\, u_2(2)\, \ldots\, u_N(N) \tag{6-47}$$

Application of the N-electron antisymmetrizer to this product function yields a function that consists of N! products combined together, as there are N! products of permutation operators in Eq. (6-45a). These N! terms can be obtained by expanding the following determinant.

$$\psi = \frac{1}{\sqrt{N!}} \begin{vmatrix} u_1(1) & u_2(1) & u_3(1) & \ldots & u_N(1) \\ u_1(2) & u_2(2) & u_3(2) & \ldots & u_N(2) \\ u_1(3) & u_2(3) & u_3(3) & \ldots & u_N(3) \\ \ldots & \ldots & \ldots & \ldots & \ldots \\ u_1(N) & u_2(N) & u_3(N) & \ldots & u_N(N) \end{vmatrix} \tag{6-48}$$

The diagonal of this determinant follows the product of Eq. (6-47). Each column uses a different spin-orbital function, and each row uses a different electron's coordinates. $\psi$ is called a *Slater determinant* after the inventor of this device, and we have that $\psi = A_N\Phi$.

Slater determinants are functions that are immediately seen to be properly antisymmetric with respect to exchange of the coordinates of any pair of electrons. Such an exchange would correspond to interchanging two rows of the Slater determinant, and it is a property of determinants that their value changes sign if two rows are exchanged. This means that the expanded form of the determinant, which is $\psi$, will have an opposite sign when a pair of electron coordinates are exchanged, and that comes about by the corresponding exchange of rows of the determinant.

There is another important property revealed by the Slater determinant construction. If there are two identical spin-orbitals, then there will be two identical columns in the Slater determinant. It is a property of determinants that their value is zero if two columns are identical. Thus, if we consider two different electrons to be in the same spin-orbital in an original product function [Eq. (6-47)], then antisymmetrization (e.g., the construction of the corresponding Slater determinant) will produce zero: Such a wavefunction is not permitted. Thus, antisymmetrization leads to a requirement that only one electron occupy a particular spin-orbital. This is a way of stating the *Pauli exclusion principle*.

## 6.9 THE MOLECULAR ORBITAL PICTURE

The product form of an electronic wavefunction can only come about through an approximation. The electronic Hamiltonian is not separable into independent terms for different electrons. The electron-electron repulsion operator prevents this separation, and in turn, an exactly determined wavefunction cannot be formed as a product of one-electron functions, e.g., $u_1(\vec{r}_1)u_2(\vec{r}_2)u_3(\vec{r}_3)...$, which is the orbital form. The approximation that leads to the orbital form is an approximation to the electron-electron repulsion. The physical aspect of the approximation is to replace the electron-electron repulsion by an electric field due to all the electrons residing in their particular spin orbitals. Then, each electron looks like its own quantum mechanical system where the potential is that of the electron's charge interacting with the positively charged nuclei and with a fixed field arising from the charge cloud of the electrons in the molecule or atom.

The approximation to the electron-electron repulsion has a mathematical effect of converting an N-electron problem, that of the original electronic Schrödinger equation, to N one-electron problems. The approximation leads to an effective one-electron Hamiltonian, called the Fock operator.

$$\hat{F} = \hat{h} + \hat{g} \tag{6-49}$$

is the operator corresponding to an electron's kinetic energy and its attraction for the nuclei.

$$\hat{h} = -\frac{\hbar^2}{2m_e}\nabla^2 - \sum_{\alpha}^{\text{nuclei}} \frac{Z_\alpha e^2}{r_\alpha} \tag{6-50}$$

$r_\alpha$ is the distance between the a nucleus and the electron. In Eq. (6-49), $\hat{g}$ is the operator associated with the effective field of all the electrons. Of course, an electron cannot be said to interact with itself, and so we really want a field operator for the field of all the *other* electrons. But, it turns out that $\hat{g}$ can be constructed so that upon applying the operator, the self-term, meaning the part for the interaction of the electron with its own share of the field, vanishes. As a result, the field operator $\hat{g}$ is the same for every electron.

The orbital wavefunctions for the atom or molecule are the eigenfunctions of the Fock operator. It is, in effect, the Hamiltonian for the system, though in the sense of a Hamiltonian for one electron at a time.

$$\hat{F} u_i = \varepsilon_i u_i \tag{6-51}$$

The eigenfunctions, $u_i$, are functions of only one coordinate. They are orbitals, and just as in many other Schrödinger equations (e.g., the harmonic oscillator), there may be many different solutions. The eigenvalues associated with the orbitals are labelled by the same index, and they are referred to as *orbital energies.*

Separability of a Schrödinger equation comes about if there are independent additive pieces of the Hamiltonian. The separability implicit in Eq. (6-51) means that the approximation corresponds to a many-electron

Hamiltonian that is simply the sum of Fock operators for each electron, and we will designate this Hamiltonian as $\hat{H}_o$.

$$\hat{H}_o = \sum_{v}^{N} \hat{F}_v \tag{6-52}$$

The Fock operators in Eq. (6-52) are subscripted to indicate that each acts on independent position coordinates for the N electrons. Separability also means that the energy is a sum, and in this case, we have

$$E_o = \sum_{i}^{N} \varepsilon_i \tag{6-53}$$

$E_o$ is the eigenenergy of $\hat{H}_o$.

The field that electrons experience in this approximation can only be prescribed when the orbital wavefunctions are known. In other words, Eq. (6-51) is quite different from the other eigenequations we have considered since the operator is dependent on the solutions. We need to know the orbitals that the electrons occupy in order to know $\hat{g}$. The dilemma is that we need to know $\hat{g}$ in order to find the orbitals. Actually, this problem can be solved by a bootstrap procedure. From a set of guess orbitals, a corresponding $\hat{g}$ operator is formed for use in Eq. (6-51). If the orbitals that are then obtained are not the same as the guess, they are used in constructing a new $\hat{g}$ operator. The whole process is repeated again and again until the orbitals used to construct $\hat{g}$ turn out to be the same as the eigenfunctions of Eq. (6-51). This means that $\hat{g}$, which represents the effective field, is not prescribed from the outset but is determined in a *self-consistent* manner. The effective field, then, is usually called a *self-consistent field*.

The orbital energies provide a framework for chemical energetics. For instance, since the net energy needed to form a molecule from separated nuclei and electrons is (approximately) $E_o$, then a reaction energy may be taken to be a difference in electronic energies of reactants and products. An example is the energy for the reaction of methylene and hydrogen to produce methane, $CH_2 + H_2 \rightarrow CH_4$. Were a quantum mechanical calculation of the orbital energies of methane, methylene, and hydrogen to give the following energies,

$$E_o(CH_4) = -27.568 \text{ a.u.}$$

$$E_o(CH_2) = -26.216 \text{ a.u.}$$
$$E_o(H_2) = -1.187 \text{ a.u.}$$

then an appropriate sum and difference could be related to the reaction energy:

$$E_{rxn} = -27.568 - (-26.216 - 1.187) = -0.165 \text{ a.u.} = -104 \text{ kcal/mol}$$

In this way, orbital energies offer a simple way of estimating reaction energies.

Another energetic feature of a molecule, the *ionization potential* (IP), is the energy required to remove an electron. This is simply an orbital energy according to Eq. (6-53). (This is within the approximation of a self-consistent field and the approximation that the orbitals for the original molecule and the ionized molecule are the same.) *Photoelectron spectroscopy* (PES) is an experiment where molecules are irradiated with high-energy photons of fixed energy (i.e., monochromatic radiation). When the photon energies are greater than an ionization potential of the sample, an electron may be ejected, and its kinetic energy will be the difference between the photon energy and the ionization potential. The ejected electrons are energy-analyzed in a photoelectron experiment, and the spectrum of energies represents the body of data obtained. Electrons may be ejected from any orbital, and so the photoelectron spectrum has sharp peaks at energies usually associated with ionization from the different orbitals. Far ultraviolet radiation is used for valence orbitals, whereas X-ray radiation is used for core orbitals. The former type of experiment is designated UPS and the latter XPS.

Molecular orbital pictures have immense conceptual and qualitative value in chemistry. In many cases, this comes about through semi-quantitative use of orbital energies and qualitative thinking about the orbital shapes. The principle of forming *linear combinations of atomic orbitals* (LCAO) to construct molecular orbitals is one important aspect of molecular orbital pictures. The basic idea is to consider orbitals of two interacting fragments of a molecule as independent electron problems. An orbital from fragment A may interact with an orbital from fragment B to produce two new orbitals (one-electron states). The interaction is the element of the Hamiltonian for the system that involves both fragments, and that means it is something that must approach zero as the fragments are pulled further and further apart. The interaction between the parts is taken as a perturbation, and the problem is analyzed with second-order perturbation theory.

The first case for LCAO is the $H_2$ molecule. At infinite separation of the two atoms, the Hamiltonian is a sum of independent hydrogen atom Hamiltonians, as in

$$\hat{H}_o = \hat{H}^{(A)} + \hat{H}^{(B)}$$

where one hydrogen atom is A and the other is B. The interaction that develops when the atoms approach each other is an attraction of each atom's electron for the other's proton plus the repulsion between the two electrons. This interaction is a perturbation, $\hat{H}_1$. There is also a repulsion between the two nuclei, and within the Born-Oppenheimer approximation, this is a constant that is added to the electronic energy at each (fixed) internuclear separation.

In the absence of the perturbation, the energies of the two electrons are the hydrogen atom energies, and it is common to represent this zero-order situation by two horizontal lines alongside a vertical (orbital) energy scale. Since these two lines place the true energies of the atoms at a large separation, one line is drawn on the far left and the other on the far right, as in Fig. 6.7. In the presence of the perturbation the orbitals of the left atom and the right atom may mix with each other. The qualitative form of this mixing is an additive combination, one plus the other, and the opposite, one minus the other. At the lowest order of perturbation theory that yields a nonzero effect of $\hat{H}_1$, one of the new orbitals is raised in energy, and the other is lowered by a like amount. The one that is raised is an unfavorable LCAO, and it is the one that has a node between the two atoms. It is called an *antibonding orbital*. The one that is lowered is more stable than the original orbitals, and so it is a bonding orbital. Since $H_2$ has two electrons, they will occupy this orbital in the $H_2$ ground electronic state. Of course, there are excited electronic states from the promotion of one or both electrons to the higher lying orbital.

In a more complicated diatomic, the orbitals of atom (fragment) A are paired up with the orbitals of atom (fragment) B on the basis of similar energies, and the same analysis is carried out. Implicit in this picture is that electrons are interacting largely one on one with electrons in the other fragment. This is partly because with low-order perturbation theory, the

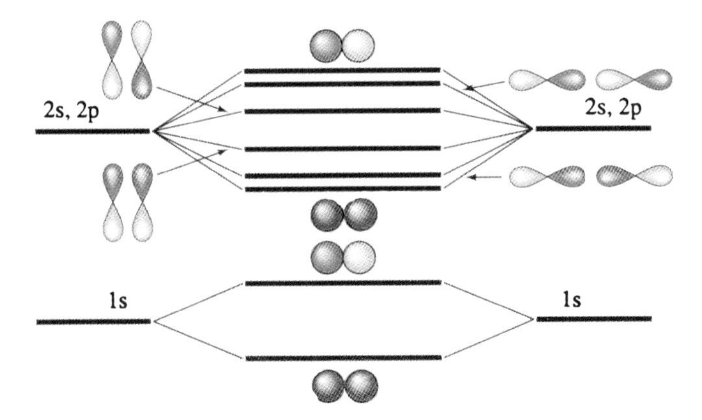


**FIGURE 6.7** An orbital energy correlation diagram for the linear combination of atomic orbitals to form molecular orbitals. The vertical axis is an orbital energy scale, and the leftmost and rightmost horizontal lines represent the orbital energies of two noninteracting hydrogen atoms. In the middle of the figure are two horizontal lines representing the energies of mixed functions of the left and right hydrogen 1s orbitals. Their energies are qualitatively deduced from first-order degenerate perturbation theory, which requires that the two mixed states be above and below the energies of the unmixed, degenerate states by equal amounts. The qualitative form of the mixed orbitals is also shown.

energetic effects will be small if the orbital energies of the two electrons from the fragments are much different in energy. (A large energy difference would lead to a sizable energy denominator in the perturbation theory expressions for the energy and wavefunction corrections.) Fig. 6.8 puts this into practice for the CO molecule. The orbital energies of carbon are represented by the series of horizontal lines on the left, and the orbital energies of oxygen are on the right. These orbitals correlate with the molecular orbitals whose approximate energies are represented by the horizontal lines in the center. The correlation of atomic orbitals with the molecular orbitals is indicated by dashed lines. Notice that the oxygen 1s orbital is not paired with a carbon orbital. Its orbital energy is much lower than any of the carbon atom's orbitals, and so, from low order perturbation theory, little mixing is expected with the carbon orbitals. This is in keeping with not considering an oxygen 1s orbital to be a valence orbital. It is an orbital that closely surrounds the oxygen nucleus, and we should expect it to be little affected by chemical bonding.

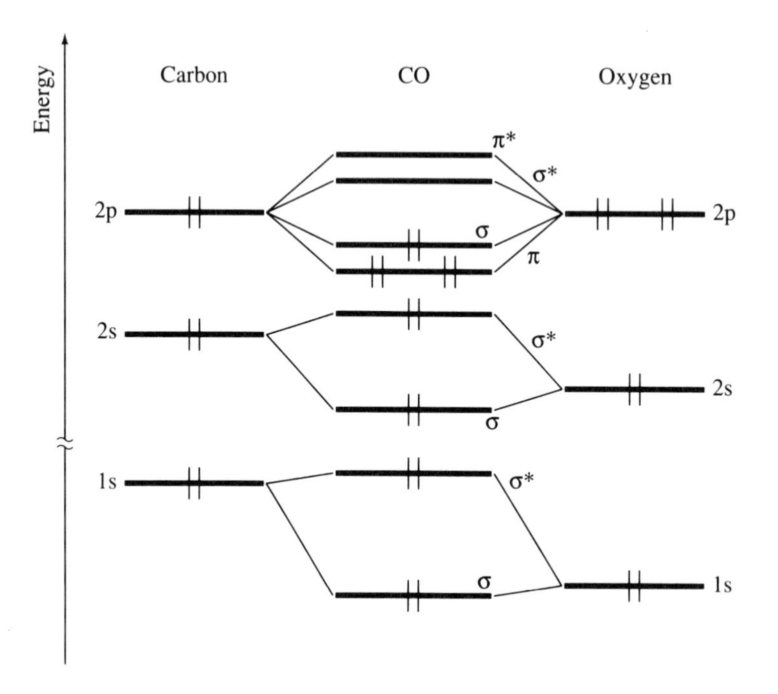


**FIGURE 6.8** An orbital correlation diagram for carbon monoxide. The carbon atomic orbital energies are on the left, and the oxygen atomic orbital energies are on the right. The molecular orbitals that form from mixing of the atomic orbitals are represented by the horizontal lines in the center at their approximate orbital energies in the CO molecule. The vertical lines indicate the orbital occupancy.

The correlation of atomic orbitals with molecular orbitals, as in Fig. 6.8, ranks the molecular orbitals in terms of their expected energies. The lowest energy electronic state of a molecule is expected to arise when the electrons fill up the molecular orbitals from the bottom up; that is, they occupy the most stable orbitals. A chemical bond is said to exist when two electrons occupy a bonding orbital, which is one that does not have a nodal plane between the two atoms. $H_2$ has one bond. When antibonding orbitals are occupied, the net bonding is taken to be given by a number called the *bond order* (BO):

$$\text{Bond order} = (\text{no. of electrons in bonding orbitals} - \text{no. of electrons in anti-bonding orbitals}) / 2$$

Thus, $He_2$ has a bond order of zero because, according to Fig. 6.7, two electrons would be in a bonding orbital just as in $H_2$, but two electrons would be in an antibonding orbital as well. Carbon monoxide has a bond order of 3. The bond order gives a rough idea of the bond strength, or of the relative bond strengths.

## 6.10 VISIBLE-ULTRAVIOLET SPECTRA OF MOLECULES

The term "spectroscopic states of molecules" refers to the states that are involved in transitions seen by spectroscopic measurement. For molecules containing atoms of other than the very heavy elements, these states have specific net electron spin. They are singlet states, meaning the spin quantum number, S, is zero, or they are doublet states ($S = 1/2$), or triplet states ($S = 1$), or states with a still greater spin quantum number. They will also have certain properties that reflect symmetry in the molecule, if any.

Visible-ultraviolet spectroscopy of molecules examines the nature of excited electronic states, since typically the absorption of a photon in this energy regime electronically excites the molecule. Probably the strongest selection rule is

$$\Delta S = 0 \tag{6-54}$$

That is, transitions are normally between states of the same spin. Other selection rules relate to the geometrical symmetry of the molecule. The molecular transitions seen in the visible and ultraviolet regions of the spectrum must be transitions from one rotational-vibrational-electronic state to another. We shall consider this in detail for a hypothetical diatomic molecule for which the ground state and excited state potential curves are those shown in Fig. 6.9. For both electronic state potential energy curves there will be a set of vibrational states and rotational sublevels. Notice that the equilibrium distance is not the same for both curves and that the curvature (i.e., the force constant) is not the same either. Thus, there will be a different vibrational frequency and a different rotational constant for

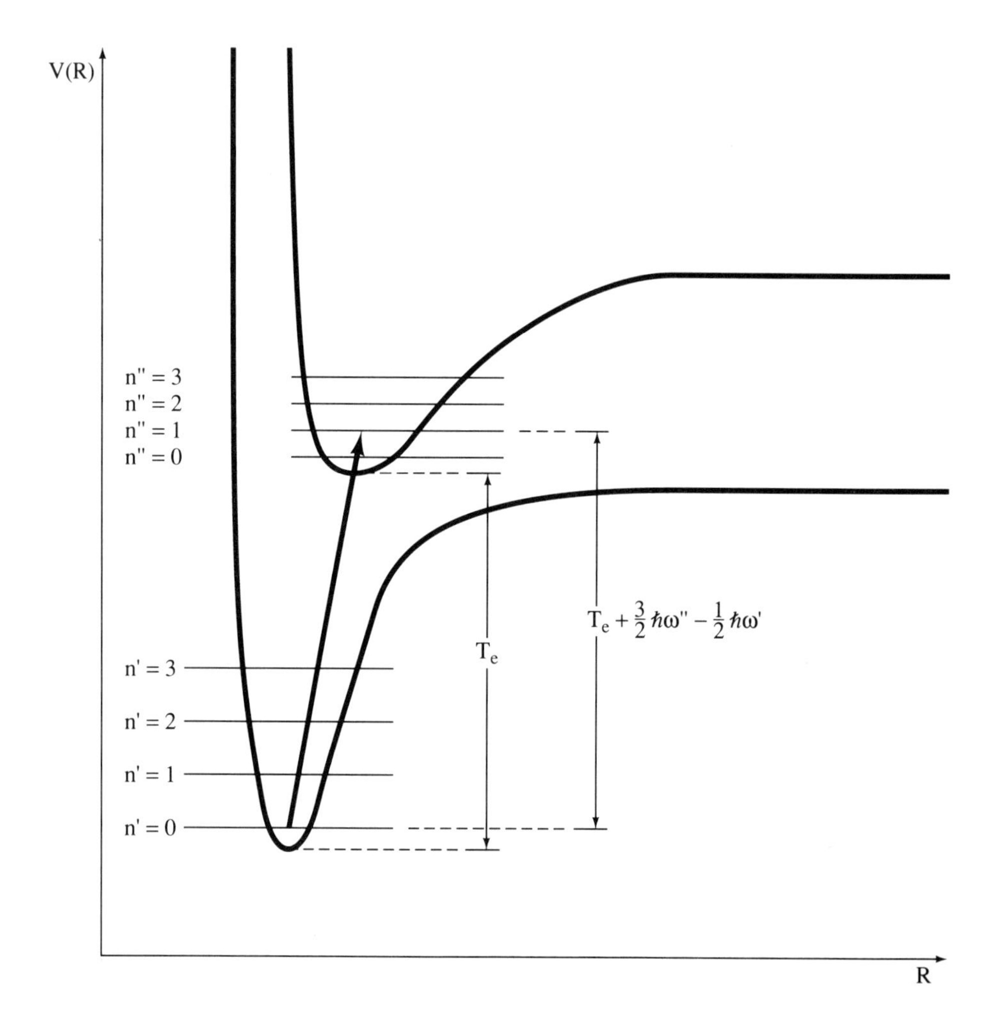


**FIGURE 6.9** Potential energy curves for two electronic states, the ground state and an excited state of the same spin, of a hypothetical diatomic molecule. Each potential may be analyzed independently to yield vibrational-rotational states and energy levels. The lowest four vibrational levels are shown. In an absorption spectrum transitions may originate from the vibrational-rotational levels of the ground electronic state and end in the vibrational-rotational levels of the excited state following the appropriate selection rules. The energies of the transitions will be sums of the energy difference between the bottoms of the potential wells, designated $T_e$, and the difference between the vibrational-rotational state energies within their respective potentials. One such transition is drawn, and the contributions to its transition energy are depicted to the right.

each electronic state. This will have to be taken into account in working out the transition frequencies.

The practice we have followed in understanding molecular spectroscopy has been to work backward, in some sense, so as to predict the frequencies of transition lines that presumably were measured. The goal is always to assign the lines to specific transitions. This enables one to use the measured frequencies to extract the true values for the vibrational frequency, the rotational constant, and other spectroscopic constants. The same procedure holds for electronic spectra.

As seen from Fig. 6.9, a transition energy will depend first on the energy difference between the potential minima of the two states, and this is called the *term value*, $T_e$. This is the contribution to the transition energy that involves only the electronic energies. The second component of the transition energies is that due to the vibration-rotation state energies within each of the two wells. The energies of those states will depend on vibrational and rotational quantum numbers n' and J' for the initial state, and n" and J" for the final state. The simplest prediction of the pattern of transition energies that we might use to assign lines would come from using the lowest order treatment for vibrational-rotational levels of a diatomic. At this level, anharmonicity, centrifugal distortion, and vibration-rotation coupling are neglected, and so we may develop a concise expression for the transition energies.

$$\Delta E = E_{final} - E_{initial} = E''(n'', J'') - E'(n', J') \qquad (6\text{-}55a)$$

$$\Delta E = T_e + (n'' + \frac{1}{2})\hbar\omega'' + B'' J''(J''+1) - (n' + \frac{1}{2})\hbar\omega' - B' J'(J'+1) \qquad (6\text{-}55b)$$

The selection rules for J in the case of an electronic transition of a diatomic molecule is that J can change by one or can be unchanged. This gives three cases for the part of the transition energy in Eq. (6-55) associated with the rotational energy, which is $\Delta E^{rot}$.

$$\Delta E = T_e + (n'' + \frac{1}{2})\hbar\omega'' - (n' + \frac{1}{2})\hbar\omega' + \Delta E^{rot}$$

$$\Delta J = J'' - J' = 1 \quad \Rightarrow \quad \Delta E^{rot} = B''(J'+1)(J'+2) - B'J'(J'+1) \qquad (6\text{-}56a)$$

$$= (B'' - B')\,J'^2 + (3B'' - B')\,J' + 2B''$$

$$\Delta J = J'' - J' = 0 \Rightarrow \quad \Delta E^{rot} = (B'' - B')\,J'^2 + (B'' - B') \qquad (6\text{-}56b)$$

$$\Delta J = J'' - J' = -1 \Rightarrow \quad \Delta E^{rot} = (B'' - B')\,J'^2 - (B'' + B')\,J' \qquad (6\text{-}56c)$$

These transition frequencies correspond to a whole series of lines for each possible n'-to-n" vibrational level changes. From the relative sizes of typical rotational constants, we should expect that $\Delta E^{rot}$ will separate the lines by a small amount in relation to the size of $T_e$. In other words, the spectrum should show a bunch of lines clustered about a transition energy of

$$T_e + (n'' + \frac{1}{2})\,\hbar\,\omega'' - (n' + \frac{1}{2})\,\hbar\,\omega'$$

This feature is called a *vibrational band* of the electronic spectrum. The rotational contributions in Eq. (6-56) allow us to decipher the *rotational fine structure* of the band.

If $B'' \approx B'$, then Eq. (6-56a) will simplify to $2B'(J'+1)$, and this is the usual expression for the R-branch lines in the fine structure of a diatomic IR absorption spectrum. Eq. (6-56b) yields zero for $\Delta E^{rot}$, and so all the lines will be at the same frequency, and they comprise a Q-branch. Eq. (6-56c) yields $-2B'\,J'$, and so these lines will correspond to the P-branch. Generally, B" and B' will be somewhat different, as suggested by the different equilibrium distances for the two potentials in Fig. 6.9. If $B'' > B'$, then with increasing J', the P- and R-branches will extend to higher and to lower frequencies, respectively, and the Q-branch lines will spread apart toward higher frequencies. If $B'' < B'$, the Q-branch lines will be spread apart but will be lower in frequency with increasing J'.

There is no strict general selection rule for the vibrational quantum number in the electronic spectra of a diatomic molecule. Transitions may be seen with n' = n" and with n' different from n" by 1, 2, 3, and so on, and these give the different bands that are observed. There will, however, be differences in the relative intensities of the bands. This is particularly apparent in low-resolution spectra, where the rotational fine structure is not resolved. The intensity differences are understood in terms of the *Franck-Condon overlap* between the initial vibrational state and the final vibrational state. Consider two electronic states with the same force constant and the same equilibrium bond length. Vibrational analysis of each potential would yield essentially the same wavefunctions for the n' = 0 state and the n" = 0 state. The overlap of these two wavefunctions would be nearly unity. On the other hand, the overlap of the n' = 0 wavefunction and the n" = 1 wavefunction would be near zero, just as the overlap of the n' = 0 and n' = 1 wavefunctions would be zero. If the equilibrium length of the excited state were made greater, the overlap between the n' = 0 and n" = 0 wavefunctions would diminish, while the overlap between the n' = 0 and n" = 1 wavefunction might increase. This makes sense from a physical standpoint: With the same equilibrium lengths, the transition is favored because in both states the molecule vibrates about the same equilibrium point. With different equilibrium lengths, the vibrational motion has to make a more drastic change upon the electronic excitation of the molecule, and so the likelihood of the transition is diminished.

Changes in the curvature (force constant and vibrational frequency) of one potential relative to the other can also change the overlap. In a rough way, the intensities of the bands will depend on the overlap between the initial and final vibrational states, and that overlap will depend on the nature of the two potential curves.

*Hot bands* are transitions from excited vibrational levels of the ground electronic state, i.e., n' > 0. They may appear at lower transition frequencies than the transition from n' = 0 to n" = 0.

Gas phase electronic spectra of polyatomic molecules are more complicated than the spectra of diatomics. The number of vibrational modes and the possibility of combination bands will usually lead to numerous vibrational bands, and these may be overlapping. Also, the rotational fine structure tends to be more complicated, as we might expect from the differences between diatomic and polyatomic IR spectra. Conventional absorption spectra can prove to be a difficult means of measuring and assigning transitions, and so numerous experimental methods have been devised to select molecules in specific initial states and probe the absorption or the emission spectrum with narrow frequency range lasers.

## 6.11 ELECTRON CORRELATION AND BOND BREAKING

The self-consistent field approximation provides a very powerful means of working with electronic wavefunctions. Perhaps most important is that it is the basis for the orbital description of molecules, and orbitals have proven quite useful in interpetation of molecular bonding and chemical phenomena. Even so, the SCF approximation is but an approximation, and it does leave errors in the energetics and the charge densities. For instance, it is a typical feature of an SCF wavefunction that electron density is slightly exaggerated in bonding regions, or regions between two nuclei. This is because the self-consistent field is an on-average description of how electrons interact, whereas their instantaneous interactions as real particles – not charge clouds – is significant. This instantaneous interaction is called *electron correlation*, and electron correlation effects are generally any effects from going beyond the SCF approximation.

Electron correlation effects may be inferred from experiment but cannot be directly measured. In the laboratory, it is not possible to turn on and turn off the self-consistent field approximation so as to see the differences. Experimental information on correlation effects is usually of the sort where a sharp qualitative difference with an orbital description is detected. This would happen in the case of a serious breakdown of the SCF approximation.

Subtle electron correlation effects are most often understood on the basis of detailed quantum chemical calculations rather than experiment. Wavefunctions and energies are obtained first from SCF calculations and then from treatments that go beyond SCF. The differences are the correlation effects. The exaggeration of electron density in bonding regions, for instance, is corrected as electron correlation is taken into account. As a result, there is often a net effect of electron correlation that amounts to making a bond length somewhat longer and to making the bond less stiff, which means diminishing the force constant. For small covalent molecules, high-level treatment of electron correlation shows bond length changes that are mostly 0.01 to 0.03 Å and reduction of harmonic force constants by about 5 to 20%.

TABLE 6.4 Comparison of Calculated and Measured Values of Certain Molecular Properties. [a]

| | Value From SCF | Value With Correlation | Experiment |
|---|---|---|---|
| LiH equilibrium bond length | 1.6071 Å | 1.5974 Å | 1.5957 Å |
| LiH n = 0 to n = 1 vibrational transition frequency | 1389.5 $cm^{-1}$ | 1358.1 $cm^{-1}$ | 1359.8 $cm^{-1}$ |
| HF n = 0 to n = 1 vibrational transition frequency | 4324 $cm^{-1}$ | 3977 $cm^{-1}$ | 3961 $cm^{-1}$ |
| $H_2O$ harmonic vibrational frequencies | 4132 $cm^{-1}$ | 3830 $cm^{-1}$ | 3832 $cm^{-1}$ |
| | 1771 $cm^{-1}$ | 1677 $cm^{-1}$ | 1649 $cm^{-1}$ |
| | 4236 $cm^{-1}$ | 3940 $cm^{-1}$ | 3943 $cm^{-1}$ |
| $N_2$ quadrupole electrical moment | 0.886 a.u. | 1.085 a.u. | 1.09 a.u. |

[a] These values are collected in Ref. 6 in the bibliography for this chapter.

Electron correlation will affect the charge distributions in ways that are manifested in the molecular properties. The dipole moment of a molecule is often changed by 5 to 10% because of electron correlation effects. A selection of calculated values for small molecules from the SCF level of treatment and from treatments that carefully include electron correlation effects is shown in Table 6.4. Comparison with measured values provides an assessment of the suitability of the SCF orbital picture in describing molecular systems.

One of the key problems with the SCF approximation is in the description of breaking, or forming, a chemical bond. In the course of breaking a bond, significant changes in the orbital character of the species are not unlikely, and even the form of the wavefunction may not be the same throughout the whole process. For instance, the orbital picture of the nitrogen molecule at equilibrium is a closed shell wavefunction, and the SCF level treatment is reasonable. It dissociates, however, into two identical atoms, and their states cannot be described as closed shells. So, a wavefunction that is appropriate throughout must include several configurations whose importance changes with the geometrical parameters. The use of several configurations means that correlation effects are being accounted for; however, these are nondynamical correlation effects, as opposed to the dynamical correlation* effects discussed in the prior section.

One powerful approach to the problem of bond breaking is the valence bond method. To understand the basic idea of the valence bond picture, let us consider the electronic structure of the ground, singlet state of the hydrogen molecule. The form of the valence bond (or Heitler-London) wavefunction is

$$\psi(1,2) = \frac{N}{\sqrt{2}} \left( a(1)\, b(2) + a(2)\, b(1) \right) \left( \alpha(1)\, \beta(2) - \alpha(2)\, \beta(1) \right) \qquad (6\text{-}57)$$

where we use the shorthand notation of electron numbers 1 and 2 in place of the spatial and spin coordinates of those electrons [e.g., a(1) instead of $a(\vec{r}_1)$ and $\alpha(2)$ instead of $\alpha(\vec{s}_2)$ ]. N is a normalization factor, the spatial functions have been designated a and b, and $\alpha$ and $\beta$ are the usual electron spin functions. This wavefunction is not a Slater determinant, but it is nonetheless properly antisymmetric with respect to particle interchange. In an SCF orbital description, the spatial functions are orthogonal. In the valence bond description, that constraint is relaxed: a and b are allowed to be the same or different spatial functions. This additional flexibility in the wavefunction provides for a better description of the changing electronic structure that occurs during bond breaking and bond formation.

When the two hydrogen atoms in $H_2$ are very far apart, perhaps 1000 Å, there is essentially no interaction. In this limiting case, the electronic structure is the electronic structure of two ground state hydrogen atoms. The wavefunction in Eq. (6-57) can represent this limiting case, which is to say that variational optimization of $\psi$ will yield the correct 1s orbitals. We can see that the wavefunction has this flexibility by replacing a with $1s_A$,

* The distinction between dynamical and nondynamical correlation is not sharp. However, as a working definition, nondynamical correlation is manifested as a few configurations entering the wavefunction with sizable importance. For dynamical correlation, there are many configurations with small expansion coefficients and a single dominant configuration.

meaning a hydrogen 1s orbital on hydrogen A, and by replacing b with $1s_B$, where B is the other hydrogen.

$$\psi(1,2) = \frac{N}{\sqrt{2}}\left(1s_A(1)\,1s_B(2) + 1s_A(2)\,1s_B(1)\right)\left(\alpha(1)\,\beta(2) - \alpha(2)\,\beta(1)\right)$$

$$= \frac{N}{\sqrt{2}}\left\{\,|1s_A\,\alpha \quad 1s_B\,\beta\,| - |1s_A\,\beta \quad 1s_B\alpha\,|\right\} \qquad (6\text{-}58)$$

Since the orbitals $1s_A$ and $1s_B$ are orthogonal, given the large A-B separation distance, the rearrangement of the wavefunction into two Slater determinants in Eq. (6-58) allows us to conclude that this is the form of the wavefunction for two unpaired electrons with singlet coupled spins. In other words, the wavefunction has the flexibility to take on the correct form at the separated atom limit.

The other limiting situation is the equilibrium separation of the protons in the $H_2$ molecule. The molecular orbital SCF picture places the two electrons in one spatial orbital, $\sigma$. We may see that the valence bond wavefunction may take on this form by letting both orbitals a and b in Eq. (6-57) be $\sigma$.

$$\psi(1,2) = \sqrt{2}\,N\,\sigma(1)\,\sigma(2)\left(\alpha(1)\,\beta(2) - \alpha(2)\,\beta(1)\right) \qquad (6\text{-}59)$$

This is recognized to be the closed-shell SCF determinant written in expanded form.

Showing that the wavefunction has the flexibility to properly describe the separated atoms and the bonded atoms means that a variational determination of the orbitals a and b at each separation distance will provide an appropriate description of the system continuously from one limiting situation to the other. In contrast, the SCF orbital picture is not uniformly appropriate for the breaking of the $H_2$ bond. We may see this by viewing the SCF $\sigma$ orbital to be essentially a linear combination of the two hydrogenic 1s orbitals:

$$\sigma = n\left(1s_A + 1s_B\right) \qquad (6\text{-}60)$$

where n is a factor to ensure normalization of the $\sigma$ orbital. If this is substituted into the SCF wavefunction, which is given in Eq. (6-59), the following is obtained.

$$\psi = N'\left(1s_A(1)\,1s_A(2) + 1s_A(1)\,1s_B(2) + 1s_B(1)\,1s_A(2) + 1s_B(1)\,1s_B(2)\right)$$

$$\times\ \left(\alpha(1)\,\beta(2) - \alpha(2)\,\beta(1)\right) \tag{6-61}$$

where all the normalization factors have been collected into N'. This is the form of the SCF wavefunction everywhere, at least while restricting the $\sigma$ orbital to be a linear combination of just the 1s orbitals. So, even at 1000 Å separation, it would have this form. If we compare it with the proper separated limit form of Eq. (6-58), we see a difference. The SCF wavefunction includes two additional terms, $1s_A(1)1s_A(2)$ and $1s_B(1)1s_B(2)$. These terms are interpreted as corresponding to an ionic arrangement of the electron density. The first places both electrons on hydrogen A, leaving bare the proton of hydrogen B. The second term is the opposite ionic form. The nonionic terms, as in Eq. (6-58), correspond to a covalent sharing of the electrons.

This analysis reveals that the SCF description weights the ionic and covalent terms equally for *all* A-B separation distances, whereas at the separated limit, the ionic terms must vanish. So, the flexibility in the valence bond description that comes about from allowing the two electrons of a bonding pair to be in different, nonorthogonal orbitals (i.e., a and b) translates into the flexibility to change the weighting between ionic and covalent parts of the electron distribution. In this way, the valence bond picture is extremely useful for describing bond breaking and bond formation.

With some manipulation, valence bond wavefunctions can be restated in terms of orthogonal orbitals. For the ground singlet state of $H_2$, it can be shown that a description equivalent to Eq. (6-61) is

$$\psi = c_1\,|\phi_1\alpha\ \ \phi_1\beta| + c_2\,|\phi_2\alpha\ \ \phi_2\beta| \tag{6-62}$$

This wavefunction is a linear combination of two configurations, and their expansion coefficients, $c_1$ and $c_2$, must be obtained variationally. Each configuration places both electrons in the same spatial orbital, and $\phi_1$ and $\phi_2$ are orthogonal. This representation of a valence bond wavefunction proves particularly helpful in computer calculations of the wavefunctions as in the *generalized valence bond* (GVB) method.

## Exercises

1. Find the explicit form of the associated Laguerre polynomial $L_3^2(z)$.
2. Show that $R_{30}(r)$ is normalized.
3. Insert $R_{41}(r)$ into Eq. (6-2) and verify that the eigenenergy is that expected from Eq. (6-9).
4. Find the energy of the n = 1 state of the hydrogen atom if the proton mass were infinite. This value is the Rydberg constant, $R_\infty$.
5. Find the energy of a hydrogen atom in a state with n = 100 and $l$ = 0. Also, find the expectation value of r in this state. (Make sure to use the correct reduced mass rather than $m_e$.)
6. The ionization energy of a one-electron atom is the energy required to promote the electron from n = 1 to n = ∞. Find the ionization energy in $cm^{-1}$ for the one-electron atoms, $He^+$, $Li^{2+}$, $C^{5+}$ and $Ne^{9+}$. In what regions of the electromagnetic spectrum (infrared, visible, ultraviolet, etc.) are there photons of energy sufficient to ionize the electron in these species?
7. To get an idea of the contraction in the size of the 1s orbitals with increasing nuclear charge, Z, calculate the expectation value of r for a single electron in a 1s orbital about the nucleus of each of the rare gas elements ($He^+$, $Ne^{9+}$, etc.).
8. Use the table method for electron arrangements to show that S = 0 and L = 0 for an $s^2$, $p^6$ and $d^{10}$ occupancy. Then show that the term symbols for a $1s^2\,2p^1$ occupancy are the same as the term symbols for a single electron in a 2p orbital.
9. Find the atomic term symbols for the beryllium states that would be associated with the occupancy $1s^2\,2s^1\,2p^1$.
10. Find the atomic term symbols that are associated with an atomic occupancy $1s^2\,2s^2\,2p^1\,3p^1$.
11. Find the Russell-Saunders term symbols for the possible states of the oxygen atom with occupancy $1s^2\,2s^2\,2p^4$.
12. Find the atomic term symbol for the ground state of the nitrogen atom.

13. Show that the term symbols for an occupancy $3d^9$ are the same as the term symbols for the occupancy $3d^1$. Then show that the term symbols for the occupancy $3d^8$ and $3d^2$ are the same. What does this suggest?

14. What are the term symbols for the ground state occupancies of Mn and $Mn^+$?

15. What is the expected spin multiplicity for the ground states of halogen atoms?

16. Compute the spin-orbit interaction energy in terms of the parameter $\alpha$ in Eq. (6-24) for the states of an atom with occupancy $1s^2\ 2s^2\ 2p^6\ 3s^2\ 3d^2$.

17. For some given values for the L and S quantum numbers of an atomic state, what is the largest possible spin-orbit interaction energy. Express this in terms of the parameter $\alpha$ in Eq. (6-24). [Hint: Consider how J in Eq. (6-24) is related to the L and S values and attempt to maximize $E^{spin\text{-}orbit}$.]

18. What are the allowed transitions between the ground state of the carbon atom and the excited states associated with the electron occupancy $1s^2\ 2s^1\ 2p^3$?

19. What should be the first allowed transition from ground state neon? (Identify the excited occupancy and the term symbols of the initial and final states.)

20. Find the term symbols for the states of an atom that are assoictaed with an occupancy $1s^2\ 2s^2\ 2p^2\ 3d^1$. (Hint: Though a table procedure may be used for all three p and d electrons, it is more concise to use a table for the two equivalent p electrons and then to use rules for inequivalent electrons to complete the angular momentum coupling.)

21. Make a contour plot of the function $V(x,y) = 3x^2 + y^2$ showing the contours for $V = 1.0$, $V = 2.0$, and $V = 3.0$ on an xy-grid. Do the same for the function $V'(x,y) = 3x^2 + y^2 + xy$. In what ways do the contours differ?

22. For a collection of identical spin 3/2 particles, what is the analog of the Pauli exclusion principle? In other words, for the wavefunction to be antisymmetric and to be an orbital product form, how many particles may be found in the same orbital?

23. For a system of four electrons in different spin-orbitals, apply the antisymmetrizer of Eq. (6-45a) to a simple product of the four spin

orbitals and show that this is the same function as the Slater determinant constructed from the orbital product.

24. Verify that a Slater determinant for the lithium atom corresponding to two 1s$\alpha$ electrons and one 1s$\beta$ electron vanishes.

25. Apply the antisymmetrizing operator to the simple orbital product function 1s$\alpha$ 2s$\alpha$ 3s$\alpha$, or else write out the terms of the Slater determinant with 1s$\alpha$ 2s$\alpha$ 3s$\alpha$ on the diagonal. Then apply the antisymmetrizing operator to each of the terms individually and simplify the result.

26. From qualitative LCAO arguments, predict the bond order of the following diatomics: LiH, HeBe, LiF, $C_2^{2-}$, $CO^+$ and $NeF^+$.

27. Repeat the analysis of Eq. (6-56) with a centrifugal distortion term, $D'[J'(J' + 1)]^2$, for the initial state and $D''[J''(J'' + 1)]^2$ for the final state.

28. If the plus and minus signs in the spin and spatial parts of the wavefunction in Eq. (6-61) were interchanged, what spin state would result?

## Bibliography


1. G. Herzberg, *Atomic Spectra and Atomic Structure* (Dover, New York, 1944). This detailed account of atomic structure offers many examples of spectra and atomic energy level diagrams for different elements.

2. G. W. King, *Spectroscopy and Molecular Structure* (Holt, Rinehart and Winston, New York, 1964).

3. A. C. Hurley, *Introduction to the Electron Theory of Small Molecules* (Academic Press, New York, 1976). This is an advanced level text.

4. A. Szabo and N. S. Ostlund, *Modern Quantum Chemistry: Introduction to Advanced Electronic Structure Theory* (Macmillan, New York, 1982). This is a thorough, advanced text.

5. R. S. Mulliken and W. C. Ermler, *Diatomic Molecules: Results of Ab Initio Calculations* (Academic Press, New York, 1977).

6. C. E. Dykstra, *Ab Initio Calculation of the Structures and Properties of Molecules* (Elsevier, Amsterdam, 1988).

*Chapter 7*

# Magnetic Resonance Spectroscopy

*Electrons and certain atomic nuclei possess intrinsic magnetic moments that give rise to an interaction energy with an external magnetic field. The difference between the interaction energies of the different states can be probed by magnetic resonance spectroscopy, an immensely powerful technique now used for qualitative analysis, determination of molecular structure, and the measurement of reaction rates and dynamics. Its use extends to imaging macroscopic objects, and the medical application for diagnostic work is widespread. This chapter provides the quantum mechanical foundation for magnetic resonance spectroscopy.*

## 7.1 NUCLEAR SPIN STATES

Atomic nuclei consist of neutrons and protons and have a structure that is as rich as the electronic structure of atoms and molecules. In chemistry, though, nuclear structure is often unimportant since it does not change in the course of a chemical reaction. It is normally quite appropriate to consider nuclei to be point-masses with a specific amount of positive charge. However, there are some important manifestations of nuclear structure that have been exploited in developing powerful types of molecular spectroscopies.

Protons and neutrons have an intrinsic spin and an intrinsic magnetic moment. Experiments revealed that there are just two possible orientations of their spin vectors with respect to a z-axis defined by an external field, and this means that the intrinsic spin quantum number is 1/2. In this feature, protons and neutrons are similar to electrons, and all half-integer spin particles are classified as fermions. Whereas the letter S is commonly used for the electron spin quantum number, the letter I is commonly used for nuclear spin quantum numbers. Thus, $I = 1/2$ for a proton, and the allowed values for the quantum number giving the projection on the z-axis, $M_I$, are $+1/2$ and $-1/2$.

The angular momentum coupling of the intrinsic proton and neutron spins in a heavy nucleus is a problem for nuclear structure theory, and it is beyond our area of discussion. However, we may correctly anticipate certain results on the basis of electronic structure. The first of these is that atomic nuclei may exist in different energy states. It turns out that the separation in energy between the ground state of a stable nucleus and its first excited states is usually enormous in relation to the size of chemical reaction energetics. Photons from the gamma ray region of the electromagnetic spectrum may be needed to induce transitions to excited nuclear states, and so these transitions can require hundreds and thousands of times the energy for a transition to an excited electronic state. Consequently, unless something is done to prepare a nucleus in an excited state, we may assume that all the nuclei in a molecule are in their ground state in some chemical experiment.

Another result we may anticipate from electronic structure is the possible values for the total nuclear spin quantum number. For instance, the deuteron (a proton and a neutron) consists of two particles with intrinsic spin of 1/2. Were these two particles electrons, we know that the possible values for the total spin quantum number are one and zero. It turns out for the deuteron that of the two coupling possibilities, $I = 1$ and $I = 0$, the $I = 1$ spin coupling occurs for the ground state. This is a consequence of the interactions that dictate nuclear structure. For chemical applications, the key information is that the deuteron is an I = 1 particle.

The intrinsic spins of stable nuclei have been determined experimentally, and the values have been explained with modern nuclear structure theory. So, tables such as that in Appendix IV are available for looking up the spin of a particular nucleus. The rules of angular momentum coupling are an aid in remembering the intrinsic spins of certain common nuclei. For instance, the helium nucleus, with its even number of protons and neutrons, has an integer spin, I = 0. In terms of the number of protons and neutrons, the carbon-12 nucleus is simply three helium nuclei. It must have an even-integer spin, and it also turns out to be I = 0. The carbon-13 nucleus has an extra neutron and has a half-integer spin, I = 1/2.

Nuclei with an intrinsic spin of I > 0 have an intrinsic magnetic moment. Just as with electron spin, the magnetic moment vector is proportional to the spin vector. A nuclear magnetic moment can give rise to an energetic interaction with an external magnetic field as well as with other magnetic moments in a molecule, such as that arising from electron spin.

Nuclear magnetic moments are small enough to have an almost ignorable effect on atomic and molecular electronic wavefunctions. On the other hand, the electronic structure has a measurable influence on the energies of the nuclear spin states. In this situation, it is an extremely good approximation to separate nuclear spin from the rest of a molecular wavefunction. This is because the electronic, rotational, and vibrational wavefunctions can be determined while ignoring nuclear spin, as has been done so far, and then the effect of the electrons can be incorporated as an external influence on the nuclear spin states.

The separated nuclear spin problem is a very special type of problem in quantum chemistry partly because the number of states is strictly limited. For example, a proton is a spin-1/2 particle; this means that I = 1/2 and that the spin multiplicity, $2I+1$, is 2. There are only two states, one with $M_I = 1/2$ and one with $M_I = -1/2$. Then, in the hydrogen molecule where there are two protons, the number of nuclear spin states for the molecule as a whole is the product of the spin multiplicities of the two nuclei, i.e., $2 \times 2 = 4$. Clearly, in large, complicated molecules, the number of nuclear spin states may be large, but finite.

Nuclear magnetic resonance (NMR) spectroscopy is concerned with the energies of the nuclear spin states and the transitions that are possible between different states. To work out the energies of the states, we need to understand the interactions that affect nuclear spin state energies and to develop an appropriate Hamiltonian.

Interactions with nuclear spins come about through the intrinsic magnetic moments of nuclei. The nuclear magnetic moment, $\mu$, is proportional to the nuclear spin vector, I, just as the magnetic moment of an electron is proportional to its spin vector. Instead of a fundamental development of the proportionality relationship, we follow a phenomenological approach by simply using an unknown for the proportionality constant, and for now we shall call it $\alpha$.

$$\vec{\mu} = \alpha \vec{I} / \hbar \tag{7-1}$$

As we have already seen with electrons, the interaction energy of a magnetic moment and a uniform external magnetic field, $H$, goes as the dot product of the two vectors. So, the interaction Hamiltonian for a bare nucleus experiencing an applied magnetic field is

$$\hat{H} = -\vec{\mu} \cdot \vec{H} \tag{7-2}$$

If there are several non-interacting nuclei experiencing the field, then the Hamiltonian is a sum of the interactions of each of the nuclei.

$$\hat{H} = -\sum_i \vec{\mu}_i \cdot \vec{H} \tag{7-3}$$

This is the basic form of the Hamiltonian for the nuclear spin state Schrödinger equation.

Nuclei embedded in molecular electronic charge distributions experience an externally applied magnetic field at a slightly altered strength. That is, the electronic motions tend to shield the nucleus from feeling the full strength of the field. There are also situations where the response of the electron distribution to a magnetic field amplifies the strength of the field at the nucleus. Either way, the Hamiltonian in Eq. (7-3) needs to be modified to properly represent nuclei in molecules, as opposed to bare nuclei in space. Again, we may approach this phenomenologically by inserting a correction factor in Eq. (7-3) without yet establishing the fundamental basis for the factor. So, for each different nucleus, there will be a different correction, and this may be expressed as

$$\hat{H} = -\sum_i \vec{\mu}_i \cdot \left( \mathbf{1} - \sigma_i \right) \cdot \vec{H} \tag{7-4}$$

The $\sigma_i$ are the *nuclear magnetic shielding* tensors. Since **1** is the unit or identity matrix, then the multiplication of the matrix quantity in parentheses in Eq. (7-4) by the field vector will lead to a new vector, which is the effective field at the nucleus.

A helpful simplification comes about from breaking up the shielding tensor into isotropic and anisotropic parts:

$$\begin{pmatrix} \sigma_{xx} & \sigma_{xy} & \sigma_{xz} \\ \sigma_{yx} & \sigma_{yy} & \sigma_{yz} \\ \sigma_{zx} & \sigma_{zy} & \sigma_{zz} \end{pmatrix} = \frac{1}{3}\left(\sigma_{xx} + \sigma_{yy} + \sigma_{zz}\right)\begin{pmatrix} 1 & 0 & 0 \\ 0 & 1 & 0 \\ 0 & 0 & 1 \end{pmatrix}$$

$$+ \begin{pmatrix} \frac{2}{3}\sigma_{xx} - \frac{1}{3}\sigma_{yy} - \frac{1}{3}\sigma_{zz} & \sigma_{xy} & \sigma_{xz} \\ \sigma_{yx} & \frac{2}{3}\sigma_{yy} - \frac{1}{3}\sigma_{xx} - \frac{1}{3}\sigma_{zz} & \sigma_{yz} \\ \sigma_{zx} & \sigma_{zy} & \frac{2}{3}\sigma_{zz} - \frac{1}{3}\sigma_{xx} - \frac{1}{3}\sigma_{yy} \end{pmatrix}$$

The isotropic part has equal diagonal elements, and so it has been given as a constant times the unit matrix. The anisotropic part is what remains. At this point, we will ignore the anisotropic part of the shielding tensor, which means that we will take the second matrix on the right hand side of the expression above to be zero. This may be regarded as an approximation, for now, though specific experimental conditions may offer a proper justification. Thus, the isotropic shielding becomes a simple scalar quantity,

$$\sigma^{iso} = \frac{1}{3}\left(\sigma_{xx} + \sigma_{yy} + \sigma_{zz}\right) \tag{7-5}$$

Eq. (7-4) simplifies upon neglect of the anisotropic part of the shielding to become

$$\hat{H} = -\sum_i (1 - \sigma_i^{iso})\, \vec{\mu}_i \cdot \vec{H} \tag{7-6}$$

The solutions of the Schrödinger equation with this Hamiltonian provide the basic energy level information for NMR spectroscopy.

Let us use Eq. (7-6) to construct an energy level diagram for the nuclear spin states of a somewhat exotic molecule, ethynol, HCCOH. It will have four nuclear spin states because only the protons have nonzero intrinsic spin. Only two quantum numbers are needed to distinguish these states, the $m_I$ numbers for the two particles. We will designate these as $m_{I_1}$ and $m_{I_2}$, and the states will be designated as $|m_{I_1}\ m_{I_2}>$. Using Eq. (7-1) in Eq. (7-6) and letting the applied field define or be applied along the z-axis so that $\vec{I} \cdot \vec{H} = \hat{I}_z H_z$, we have that the spin states are eigenfunctions of the Hamiltonian.

$$\hat{H}\, |m_{I_1}\ m_{I_2}> = -(1-\sigma_1)\,\alpha_1 H_z\, m_{I_1}\, |m_{I_1}\ m_{I_2}> \tag{7-7}$$

$$-(1-\sigma_2)\,\alpha_2 H_z\, m_{I_2}\, |m_{I_1}\ m_{I_2}>$$

The energies of the four states have a linear dependence on the strength of the external magnetic field, and they will separate in energy with increasing field strength. Since both magnetic nuclei are protons in this example, then $\alpha_1$ must be the same value as $\alpha_2$, and we will simply use $\alpha$ for both. A tabulation of the four states' eigenenergies from Eq. (7-7) is

| $m_{I_1}$ | $m_{I_2}$ | Energy |
|---|---|---|
| −1/2 | −1/2 | $\alpha H_z / 2 \left[ (1-\sigma_1) + (1-\sigma_2) \right]$ |
| −1/2 | 1/2 | $\alpha H_z / 2 \left[ (1-\sigma_1) - (1-\sigma_2) \right]$ |
| 1/2 | −1/2 | $\alpha H_z / 2 \left[ -(1-\sigma_1) + (1-\sigma_2) \right]$ |
| 1/2 | 1/2 | $\alpha H_z / 2 \left[ -(1-\sigma_1) - (1-\sigma_2) \right]$ |

Fig. 7.1 is an energy level diagram based on these energies.

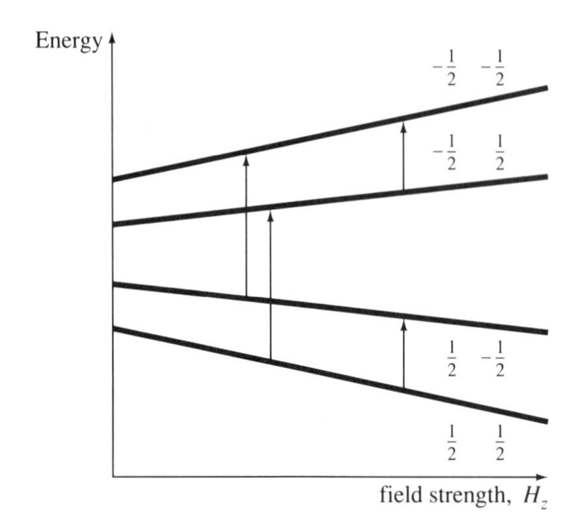


**FIGURE 7.1** Nuclear spin energy levels of a molecule such as HCCOH with two noninteracting protons in different chemical environments. The levels separate in energy with increasing strength of the external field, according to Eq. (7-6). The levels are labelled by the $m_I$ quantum numbers of the two protons. The vertical arrows indicate the allowed NMR transitions. Clearly, the energy of these transitions depends on the strength of the applied field.

In a conventional NMR experiment, transitions between the nuclear spin energy levels are induced by applying electromagnetic radiation perpendicular to the direction of the static magnetic field. The selection rules are that one $m_I$ quantum number can increase or decrease by one while all the other $m_I$ quantum numbers remain unchanged.* In effect, this says that absorption of a photon of energy will "flip" the spin of only one nucleus at a time. From the above tabulation of the energy levels for HCCOH and from Fig. 7.1, we may see that the selection rules correspond to allowed transitions from the lowest energy state to both of the next two energy states and that the transitions from these two states to the highest energy state are allowed. The arrows in Fig. 7.1 indicate these transitions.

* The interaction that leads to a transition is between the magnetic moment of the nucleus and the magnetic field of the electromagnetic radiation. These are termed "magnetic dipole transitions."

The transition energies are obtained by taking differences in the state energies. Using the tabulation above, we may obtain the following transition energies.

| Initial state | Final state | Transition energy |
|---|---|---|
| $\vert\ 1/2 \quad 1/2 >$ | $\vert\ 1/2 \ -1/2 >$ | $\alpha H_z (1 - \sigma_2)$ |
| $\vert\ 1/2 \quad 1/2 >$ | $\vert -1/2 \quad 1/2 >$ | $\alpha H_z (1 - \sigma_1)$ |
| $\vert\ 1/2 \ -1/2 >$ | $\vert -1/2 \ -1/2 >$ | $\alpha H_z (1 - \sigma_1)$ |
| $\vert -1/2 \quad 1/2 >$ | $\vert -1/2 \ -1/2 >$ | $\alpha H_z (1 - \sigma_2)$ |

This reveals that though there are four possible transitions, there are only two frequencies at which transitions will be detected. Furthermore, measuring the transition frequencies will yield values for $\alpha(1 - \sigma_1)$ and for $\alpha(1 - \sigma_2)$. The typical field strengths used in NMR are such that the transition frequencies of the electromagnetic radiation are in the microwave or radiofrequency regions of the spectrum. These are very low-energy transitions compared to vibrational or electronic excitations of a molecule. The basic experiment can be carried out in two ways. A fixed field strength can be applied and the frequency of the radiation varied until a transition is detected by a change in the power transmission of the radiation. Or, the frequency can be fixed and the power monitored as the field strength is varied. When this sweeping of the field strength brings a transition into resonance with the radiation frequency, a change in the power level of the radiation field is detected. So, transition frequencies may be measured for a given field strength, or field strengths at which transitions occur at a certain frequency may be measured. The latter can be converted to the former.

The information obtained from a low resolution NMR scan is the chemical shielding. It is usually obtained as a shift relative to some standard or reference transition. The chemical shift is designated $\delta$ and is a dimensionless quantity. It is usually a number on the order of $10^{-6}$, and so it is usually stated as being in parts per million (ppm). If the radiation frequency has been varied in the experiment, $\delta$ is given as

$$\delta \cong \frac{\omega_{ref} - \omega}{\omega_{ref}} \tag{7-8}$$

If the field strength has been swept, $\delta$ is given as

$$\delta \cong \frac{H_z - H_{z\text{-ref}}}{H_{z\text{-ref}}} \tag{7-9}$$

The chemical shifts are characteristic of the chemical environment. Thus, NMR spectra may serve as an analytical tool for determining functional groups that are present in a molecule, for instance.

Proton NMR spectra of organic molecules are usually referenced to the proton transitions of tetramethylsilane (TMS). Of course, $\delta$ of the reference, according to Eq. (7-8) or (7-9), is zero. (Another system, with values designated $\tau$, is sometimes used; the values are given as $\tau = 10 - \delta$ with $\tau$(TMS) = 10 ppm.) The NMR signature of protons in different environments in organic molecules is a transition roughly within these ranges ($\delta$ scale):

| | |
|---|---|
| Alkanes | 0 - 1 ppm |
| -C=CH- | 4 - 8 ppm |
| -C≡CH | 2 - 3 ppm |
| Aromatic | 7 - 8 ppm |
| -OH | 10 - 11 ppm |
| -CHO | 9 - 10 ppm |

More extensive lists of this sort are available.

The proportionality constant introduced in Eq. (7-1) varies from one nucleus to another because of the intrinsic nuclear structure. The proportionality constant, $\alpha$, may be replaced by a dimensionless value, g, for a given nucleus, and $\mu_o$, the nuclear magneton, or basic measure of the size of nuclear magnetic moments.

$$\vec{\mu}_i = \mu_o g_i \vec{I}_i / \hbar \tag{7-10}$$

With the external magnetic field in the z-direction, we may rewrite Eq. (7-6).

$$\hat{H} = -\mu_o H_z \sum_i g_i (1 - \sigma_i) \hat{I}_{i_z} / \hbar \tag{7-11}$$

The variation in the nuclear g values has a much more profound effect on transition frequencies than does the variation in chemical shifts for a given type of nucleus. This is evident from Eq. (7-1): A 10% variation in $\sigma$, given that $\sigma$ is on the order of $10^{-6}$, has much less effect on the energy

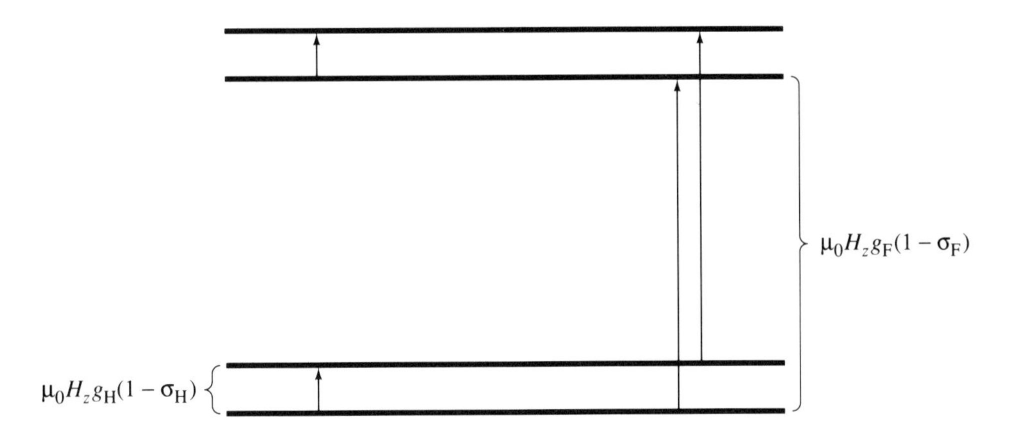


**FIGURE 7.2** An energy level diagram for the four spin states of $H^{19}F$ in an external magnetic field of fixed strength. The allowed transitions are represented by the vertical arrows. Because of the sizable difference in g values, the transition energies for the $^{19}F$ spin flip are much different than the transition energies of the proton. In practice, two different instrumental set-ups are required to observe the two transitions.

separation between states than a 10% variation in g. Consequently, NMR spectra are normally nuclei-specific; the instrumentation is set for a narrow range of frequencies to scan (or fields to sweep), and that will be for a certain type of nucleus, such as protons. A different instrumentation setup will be used for carbon-13 nuclei, or for oxygen-17, or for fluorine-19. Of course, even though we may obtain proton NMR spectra independent of obtaining carbon-13 NMR spectra, etc., the complete Hamiltonian of Eq. (7-11) is still at work and still describes the complete problem. Fig. 7.2 illustrates this for the case of the $H^{19}F$ molecule. Both nuclei are spin-1/2 particles. There are four levels, but the two allowed transition frequencies are very different because of the different g values.

As a complicated example of nuclear spin energy levels, let us consider the diimide molecule, HNNH. At this stage, we are assuming no interaction between the magnetic moments of the different nuclei, and so the Hamiltonian of Eq. (7-11) is to be used. In both the trans and cis forms of diimide, the nitrogens are in equivalent environments and the protons are in equivalent environments. Nitrogen-14 nuclei have an intrinsic spin of I = 1. The spin multiplicity is 3 since $m_I$ may be –1, 0 or 1, and so with the spin multiplicity of the protons being 2, the number of spin states is $3 \times 3 \times 2 \times 2 = 36$. The energies of these states, obtained via Eq. (7-11) are

$$E_{m_{I_{N-1}} m_{I_{N-2}} m_{I_{H-1}} m_{I_{H-2}}} = -\mu_o H_z \; g_N (1 - \sigma_N) \left( m_{I_{N-1}} + m_{I_{N-2}} \right)$$

$$-\mu_o H_z \; g_H \; (1 - \sigma_H) \left( m_{I_{H-1}} + m_{I_{H-2}} \right) \quad (7\text{-}12)$$

where the N and H subscripts indicate the nitrogen and hydrogen atoms, with 1 and 2 used to distinguish the like atoms. The selection rule applied to this problem is that one and only one of the m quantum numbers can change in a transition, and that change may be 1 or –1. This gives 84 different allowed transitions, but if we systematically consider them, we can show that there are only two different transition energies. One will show up in a nitrogen-14 NMR spectrum, and it is at $\mu_o H_z g_N (1-\sigma_N)$. The other will be in a proton NMR spectrum, and it is at $\mu_o H_z g_H (1-\sigma_H)$.

## 7.2 NUCLEAR SPIN-SPIN COUPLING

The nuclear spins of different magnetic nuclei may interact and couple much the same as electron spins couple. This will often lead to small energetic effects that tend to be noticeable under high resolution. The interaction that couples nuclear spins depends on their magnetic dipoles and their electronic environments. We may treat this interaction phenomenologically rather than attempting to analyze its fundamental basis. Since the dipole moments are proportional to the intrinsic spin vectors, we may write the spin-spin coupling interaction as proportional to the dot product of two spin vectors. So, for a system of two interacting nuclei,

$$\hat{H}' = \frac{J_{12}}{\hbar^2} \vec{I}_1 \cdot \vec{I}_2 \quad (7\text{-}13)$$

where J has been introduced as the phenomenological proportionality constant.

The spin-spin coupling interaction of Eq. (7-13) is usually a small perturbation on the energies of the nuclear spin states experiencing the external field of an NMR instrument. So, it is appropriate to treat this interaction with low-order perturbation theory, particularly first-order perturbation theory. The expression for the first-order correction to an energy is the expectation value of the perturbation. For a system of two nonzero spin nuclei, the spin states are distinguished by the two quantum numbers, $m_{I_1}$ and $m_{I_2}$. The corrections to the energies of these states are given by

$$E^{(1)}_{m_{I_1} m_{I_2}} = \frac{J_{12}}{\hbar^2} < m_{I_1} m_{I_2} \mid \vec{I}_1 \cdot \vec{I}_2 \mid m_{I_1} m_{I_2} >$$

$$= \frac{J_{12}}{\hbar^2} < m_{I_1} m_{I_2} \mid \hat{I}_{x_1} \hat{I}_{x_2} + \hat{I}_{y_1} \hat{I}_{y_2} \mid m_{I_1} m_{I_2} >$$

$$+ \frac{J_{12}}{\hbar^2} < m_{I_1} m_{I_2} \mid \hat{I}_{z_1} \hat{I}_{z_2} \mid m_{I_1} m_{I_2} > \qquad (7\text{-}14)$$

Only the last term in Eq. (7-14) is nonzero, and from evaluating this integral, we obtain a simple result.

$$E^{(1)}_{m_{I_1} m_{I_2}} = J_{12}\, m_{I_1} m_{I_2} \qquad (7\text{-}15)$$

This says that the corrections to the energy due to spin-spin coupling between two nuclei will vary with the product of the quantum numbers giving z-component projections of the spin vectors.

As the first example of the effect of spin-spin interaction on NMR spectra we will consider a system with two protons attached to adjacent atoms, such as in HCOOH. The four spin states in the absence of spin-spin interaction would have the following energies, which are the zero-order energies for the perturbative treatment of the spin-spin interaction.

$$E_{-1/2\ -1/2} = \frac{\mu_o H_z g_H}{2}\left[(1-\sigma_1) + (1-\sigma_2)\right]$$

$$E_{-1/2\ 1/2} = \frac{\mu_o H_z g_H}{2}\left[(1-\sigma_1) - (1-\sigma_2)\right]$$

$$E_{1/2\ -1/2} = \frac{\mu_o H_z g_H}{2}\left[-(1-\sigma_1) + (1-\sigma_2)\right]$$

$$E_{1/2\ 1/2} = \frac{\mu_o H_z g_H}{2}\left[-(1-\sigma_1) - (1-\sigma_2)\right]$$

The subscripts on the energies are the $m_I$ quantum numbers of the two protons. The first-order corrections, according to Eq. (7-15), are the following.

$$E^{(1)}_{-1/2\ -1/2} = \frac{J_{12}}{4}$$

$$E^{(1)}_{-1/2\ 1/2} = -\frac{J_{12}}{4}$$

$$E^{(1)}_{1/2\ -1/2} = \frac{J_{12}}{4}$$

$$E^{(1)}_{1/2\ 1/2} = \frac{J_{12}}{4}$$

Notice that the first-order corrections raise the energy of two levels and lower the energy of the other two levels. As seen in Fig. 7.3, these changes in the energy levels "split" the pairs of transitions that were at like energies. The transition energies are

$$\Delta E_a = \mu_o H_z g_H (1-\sigma_1) + J_{12}/2$$

$$\Delta E_b = \mu_o H_z g_H (1-\sigma_1) - J_{12}/2$$

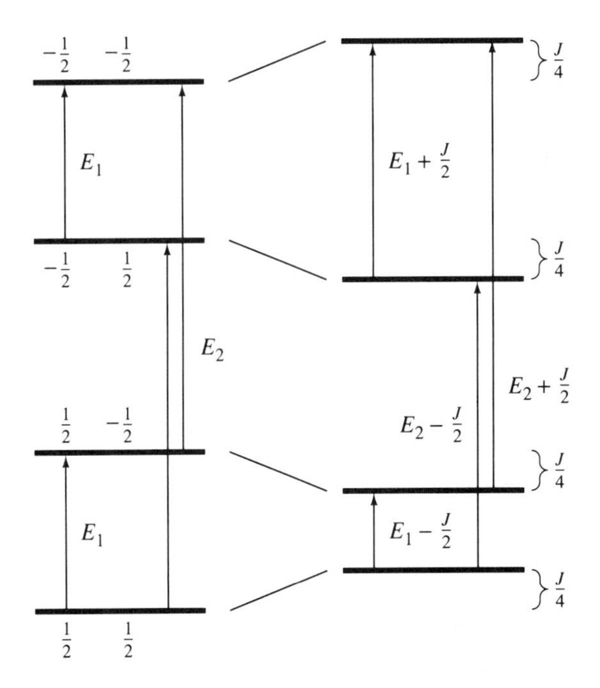


**FIGURE 7.3** Energy levels for a hypothetical molecule with two protons that are nearby but in different chemical environments. On the left are the energy levels that would be found in the absence of spin-spin interaction. On the right are the energy levels with spin-spin interaction treated via first-order perturbation theory. The vertical arrows show the allowed transitions, and the lengths of these arrows are proportional to the transition energies. Thus, the effect of the spin-spin interaction is seen to split the transitions, that is, to take each pair of transitions that would occur at the same frequency and shift one to a higher frequency and one to a lower frequency.

$$\Delta E_c = \mu_o H_z g_H (1 - \sigma_2) + J_{12}/2$$

$$\Delta E_d = \mu_o H_z g_H (1 - \sigma_2) - J_{12}/2$$

Transitions a and b are moved to higher and lower frequencies, respectively, than they would be at in the absence of spin-spin interaction. The same is true for the c and d transitions. The separation between transition a and transition b (i.e., the difference between $\Delta E_a$ and $\Delta E_b$) is

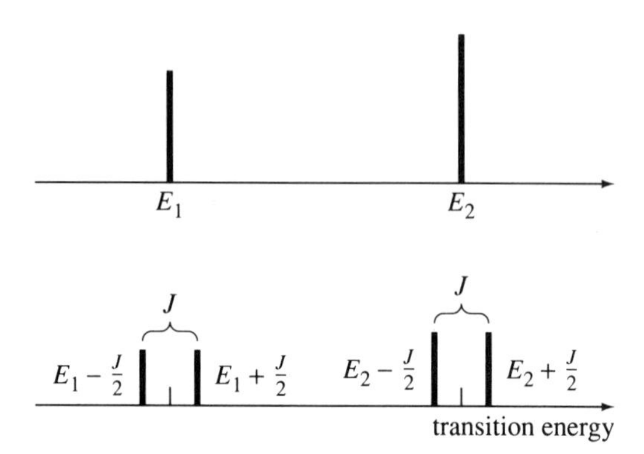


**FIGURE 7.4** Stick representation of the NMR spectra for the hypothetical molecule with the energy levels given in Fig. 7.3. The top spectrum corresponds to the energy levels in the absence of spin-spin interaction, whereas the bottom spectrum includes the effect.

---

the spin-spin coupling constant, $J_{12}$. Likewise, the separation between transition c and transition d is $J_{12}$. Fig. 7.4 is a stick representation of the spectrum that would result.

The next example is a hypothetical molecule whose magnetic nuclei are three protons in different chemical environments. The spin-spin interaction Hamiltonian for a system with many interacting nuclei is simply a sum of the pair interactions of Eq. (7-13):

$$\hat{H}' = \sum_i \sum_{j>i} \frac{J_{ij}}{\hbar^2} \vec{I}_i \cdot \vec{I}_j \qquad (7\text{-}16)$$

So, the first-order corrections to the energy are also sums of pair contributions with the form given in Eq. (7-15).

$$E^{(1)} = \sum_i \sum_{j>i} J_{ij} \, m_{I_i} \, m_{I_j} \qquad (7\text{-}17)$$

The quantum number subscripts on $E^{(1)}$ have been suppressed for conciseness; they would be the $m_I$ quantum numbers for all the nuclei. Fig. 7.5 shows how the zero-order energy levels for this hypothetical system would be affected by the first-order corrections of Eq. (7-17). The

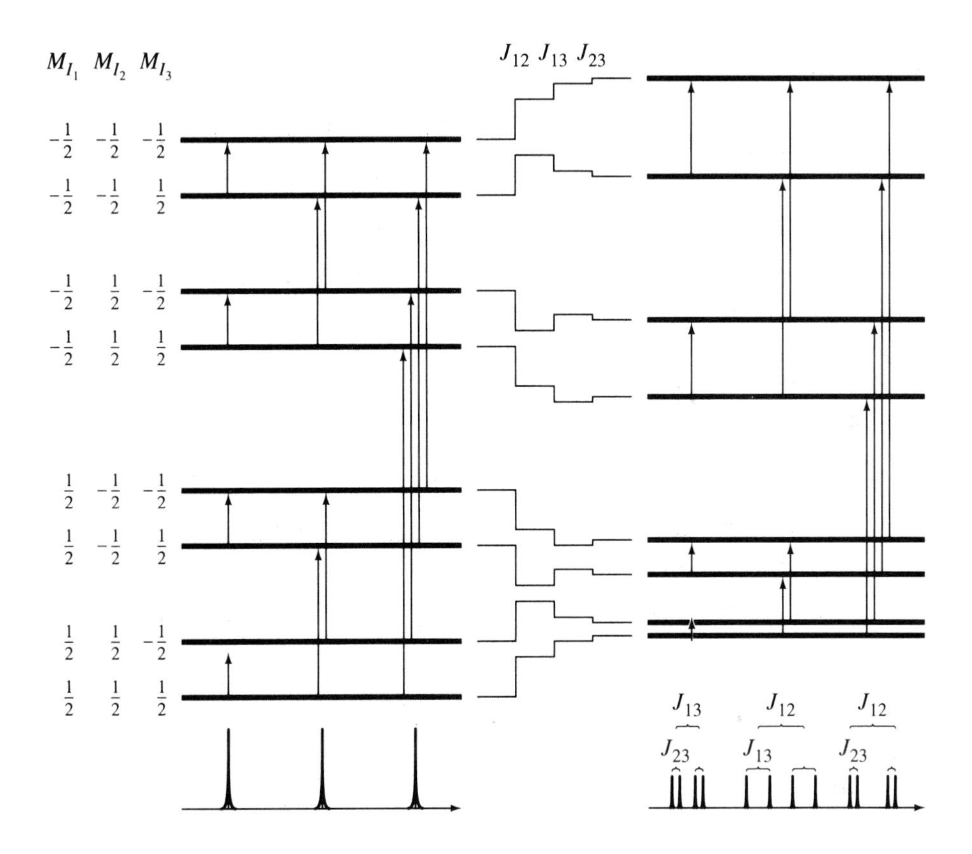


**FIGURE 7.5** Energy level diagram for the nuclear spin states of a hypothetical molecule with three interacting protons in different chemical environments. The energy levels on the left neglect spin-spin interaction. The allowed transitions give rise to a spectrum of three lines, shown as a stick representation at the bottom. These three lines correspond to a spin "flip" of each of the three different nuclei, and their relative transition frequencies give the chemical shift of each. The energy levels on the right have been obtained by first including the 1-2 coupling, then the 1-3 coupling, and finally the 2-3 coupling, assuming $J_{12} > J_{13} > J_{23}$. Transition lines are drawn for the final set of levels, and the resulting stick spectrum is shown at the bottom. The values of the three coupling constants may be obtained directly from the spectrum, as shown.

resulting spectrum consists of each original line split into a pair of lines, which are then split again. This reveals a simple rule for interpreting proton NMR spectra: A pair of closely spaced lines may have come about because of the spin-spin interaction with a single nearby proton.

Though we have treated spin-spin interaction phenomenologically (by employing the coupling constant, J, as a parameter instead of deriving it from fundamental interactions), there are certain features that we may anticipate. One of these is that the size of a coupling constant, $J_{12}$, will vary with different pairs of nuclei. It will fall off with increasing separation between the interacting nuclei to the extent that the interaction strength diminishes with increasing separation distance. In practice, proton NMR spectra generally show splitting by protons attached to adjacent atoms. So, we might expect to see effects from proton spin-spin interaction in HCOOH given the proximity of the protons, but perhaps not in HC≡CCOOH. Thus, splittings in spectra are intimately related to molecular structure and may serve to reveal structural features. Also, the lines of both of the interacting nuclei are split by the identical amount (i.e., $J_{12}$), and so finding two pairs of lines with the same splitting may identify two protons that are interacting.

Another example to consider is one with different kinds of magnetic nuclei, and a simple example is HD (deuterium-substituted $H_2$). The deuterium nucleus has a spin of one (I = 1). The $m_I$ values for the deuterium nucleus are 1, 0, and –1. The spin states will be distinguished by the $m_I$ quantum numbers for the deuterium and the proton. From Eq. (7-11) we may write the zero-order energies, and from Eq. (7-15), the first-order corrections. We may tabulate these energies and present them in a concise form by listing the possible energy terms at the head of the column, and then with each state as a row in the table, we list the factor that should be applied to the energy term.

| State: | | Energy terms: | | |
|---|---|---|---|---|
| $m_{I_D}$ | $m_{I_H}$ | $\mu_o H_z g_D (1-\sigma_D)$ | $\mu_o H_z g_H (1-\sigma_H)/2$ | $J_{12}/2$ |
| –1 | –1/2 | 1 | 1 | 1 |
| –1 | 1/2 | 1 | –1 | –1 |
| 0 | –1/2 | 0 | 1 | 0 |
| 0 | 1/2 | 0 | –1 | 0 |
| 1 | –1/2 | –1 | 1 | –1 |
| 1 | 1/2 | –1 | –1 | 1 |

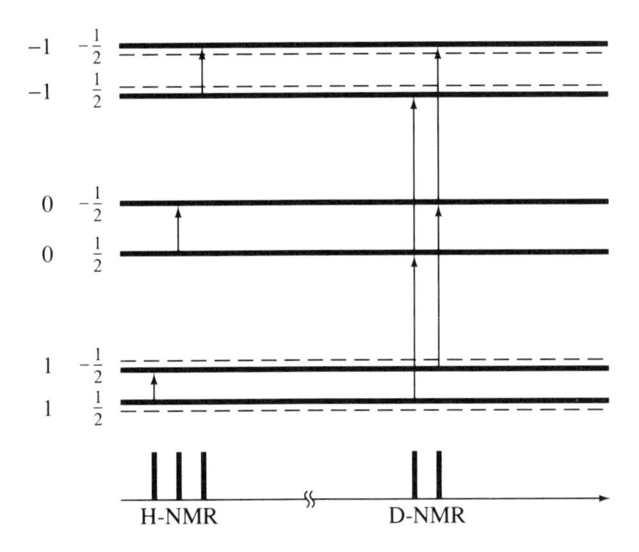


**FIGURE 7.6** Nuclear spin energy levels of the HD molecule in a magnetic field. The allowed transitions, represented by vertical arrows, are those for which the $m_I$ quantum number of the deuterium changes by one, or the $m_I$ quantum number of the proton changes by one. The deuterium and proton stick spectra are shown at the bottom.

So, the energy of the state in the second row, for instance, is the sum of the first term plus the second term multiplied by –1 plus the third term multiplied by –1. A sketch of these energy levels is given in Fig. 7.6 along with the stick spectra that are predicted. The spectra show the proton transition line split into three lines, whereas the deuterium transition is split into two lines. This reveals a generalization of the rule about nearby protons splitting lines into pairs. It is that a nearby magnetic nuclei will split a transition line into the number of lines equal to the spin multiplicity, 2I + 1, of the nucleus. The deuterium splits the proton transitions into three lines, since $2I_D + 1 = 3$.

Equivalent nuclei are the magnetic nuclei in equivalent chemical environments, such as the two protons in formaldehyde. The chemical shielding for each of these protons must be the same because of the symmetry of the molecule. The quantum mechanics for spin-spin

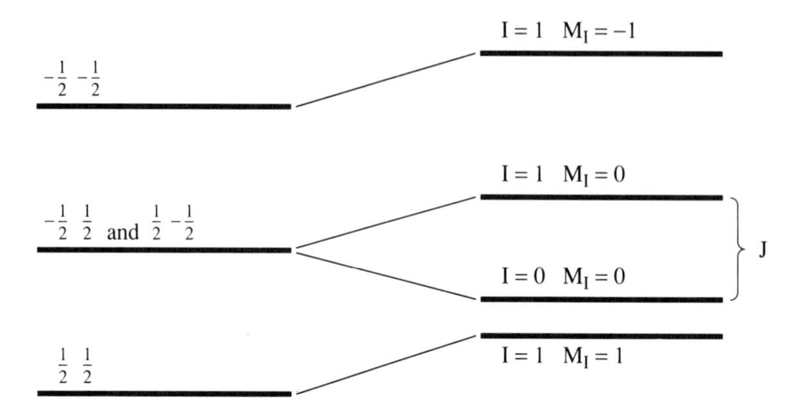


**FIGURE 7.7** Energy levels and transitions for the proton spin states of formaldehyde. On the left are the levels obtained with neglect of spin-spin interaction, and they are labelled by the $m_I$ quantum numbers of the two protons. The middle level is doubly degenerate. If spin-spin interaction is treated with first-order degenerate perturbation theory, the levels on the right result, with an assumed size of the coupling constant, J. A transition from the I = 0 state to an I = 1 state would measure J, but that is a forbidden transition. No splitting of the line in the spectrum occurs.

coupling with equivalent nuclei has some complexities that are not encountered with inequivalent nuclei, but a rather simple rule for predicting or interpreting spectra emerges just the same. The complexity of equivalent nuclei is spin state degeneracy. The zero-order energies for the proton spin states in formaldehyde with quantum numbers (1/2 −1/2) and (−1/2 1/2) are both zero according to Eq. (7-11). This is the consequence of the σ for both nuclei being the same. These two states are degenerate, and degenerate first-order perturbation theory is needed to treat the spin-spin interaction.

The result from degenerate first-order perturbation theory for two interacting, equivalent protons amounts to completely coupling their spins. Using the angular momentum coupling rules, two spin-1/2 protons may yield a net coupled spin with I = 1 or with I = 0. The selection rules for NMR transitions of equivalent nuclei turn out to be $\Delta I = 0$ and $\Delta M_I = \pm 1$. Fig. 7.7 shows that with these selection rules, it is not possible to obtain

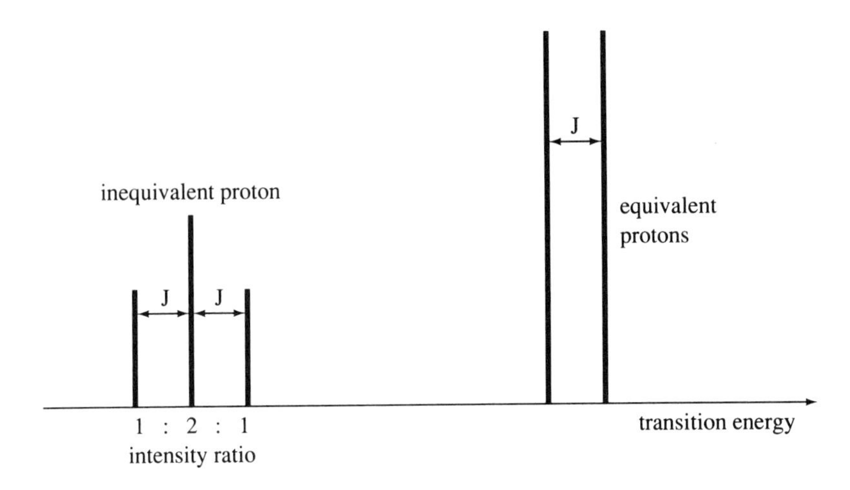


**FIGURE 7.8** The NMR stick spectrum of a hypothetical molecule with two equivalent protons and a third, nearby proton in a different chemical environment. This spectrum may be understood as a combination of the spectrum of a system with an I = 1 particle interacting with the third proton and the spectrum from an I = 0 particle interacting with the third proton. These two pseudoparticles correspond to the possible couplings of the spins of the two equivalent nuclei.

from the NMR spectrum the J coupling constant for the equivalent protons. For formaldehyde, we expect a single proton NMR transition.

The coupling of equivalent nuclei with other nuclei can produce splittings as well as important intensity features in an NMR spectrum. Consider a hypothetical molecular system with two equivalent protons and a third proton in a different environment. The NMR spectrum may be regarded as a superposition of the spectra obtained for the two coupling possibilities (namely, I = 1 and I = 0) of the equivalent protons. In the case of I = 1 coupling, we expect a spectrum that looks like a spin-one particle with a spin-1/2 particle. This is qualitatively the same as the HD molecule. In the case of I = 0 coupling, we expect a spectrum that looks like a spin-zero particle with a spin-1/2 particle. The single transition for this case will add to the transition intensity of the middle line of the triplet of lines from the first case. The result is shown in Fig. 7.8. It is that a spin flip by the third proton corresponds to three transition lines in an intensity ratio of 1 : 2 : 1. The equivalent protons' transitions are split into two lines by the third proton's interaction.

## 7.3 ELECTRON SPIN RESONANCE SPECTRA

If a molecule has a nonzero electronic spin (i.e., if S > 0), then the associated magnetic moment will interact with an external magnetic field. This is the basis for electron spin resonance (ESR) spectroscopy, which is also called electron paramagnetic resonance (EPR) spectroscopy. The electron spin energy levels are separated by the magnetic field interaction, and the resulting energy differences are probed spectroscopically. The electron spin magnetic moment will also interact with nuclear magnetic moments, and it is this complication that ends up providing much useful information from ESR spectra.

Usually, electronic orbital angular momentum is zero in polyatomic molecules, and so this source of magnetic moments can be ignored. Also, the ESR experiment is usually carried out with a liquid or solid state sample, and so there is no magnetic moment arising from molecular rotation since the molecules are not freely rotating.

The general form for the interaction Hamiltonian between an isotropically sampled magnetic field $\vec{H}$ and the electron spin is

$$\hat{H} = \frac{g_e \mu_B}{\hbar} \vec{S} \cdot \vec{H} \qquad (7\text{-}18)$$

where $\mu_B$ is the Bohr magneton. If magnetic nuclei are present, the Hamiltonian is a sum of this term and the nuclear Hamiltonian of Eq. (7-11) along with the spin-spin interaction terms. As before, we take the direction of the magnetic field to define the z-axis, and then

$$\hat{H} = \frac{g_e \mu_B H_z}{\hbar} \hat{S}_z - \frac{\mu_o H_z}{\hbar} \sum_i^{\text{nuclei}} g_i (1 - \sigma_i) \hat{I}_{i_z}$$

$$+ \sum_{i>j} \frac{J_{ij}}{\hbar^2} \vec{I}_i \cdot \vec{I}_j \quad + \sum_i \frac{a_i}{\hbar^2} \vec{I}_i \cdot \vec{S} \tag{7-19}$$

This equation introduces the constants $a_i$ phenomenologically; that is, the electron spin-nuclear spin interaction must be proportional to the dot product of the spin vectors, but the fundamental basis for the proportionality constant ($a_i$) is not considered.

The form of the Schrödinger equation that uses the Hamiltonian of Eq. (7-19) is the same as the form of the NMR Schrödinger equation. In both, there are z-component spin operators and spin-spin dot products. The only thing that is different in the ESR Hamiltonian is that one of the spins is electron spin and with it comes a different set of constants (e.g., $g_e$ versus $g_i$). Notice that $\mu_B/\mu_o$ is the ratio of the proton mass to the electronic mass (1836.15).

In NMR, the external magnetic field effect is generally very large with respect to the size of the nuclear spin-spin interaction. This is because of the need to use a strong field in order to make the transition energies great enough to place them in the radiofrequency region of the spectrum. In ESR, the much greater size of $\mu_B$ means that the choice is available to use weaker fields; at 10,000 G, ESR transitions are typically around 10,000 MHz, whereas NMR transitions are around 10 MHz. Consequently, there are two situations to consider in analyzing ESR spectra, a "high-field" case and a "low-field" case. The high-field case means that the field dominates all spin-spin interactions. The analysis, then, follows that for NMR where the spin-spin terms in the Hamiltonian are treated by first-order perturbation theory. In the low-field case, spin-spin coupling is of much greater relative importance, and first-order perturbation energies are inappropriate.

Let us use the formyl radical $HCO^{\bullet}$ as an example problem for predicting high-field and low-field ESR spectra. In this molecule, the single unpaired electron implies that there is a net electronic spin of 1/2 (i.e., S = 1/2). The proton spin, I, is 1/2, and so there are four spin states for the molecule. With high fields, the Hamiltonian is broken into a zero-order part and the spin-spin perturbation:

$$\hat{H}^{(0)} = g_e \mu_B H_z \hat{S}_z / \hbar \; - \; g_H \mu_o H_z (1-\sigma) \hat{I}_z / \hbar \tag{7-20a}$$

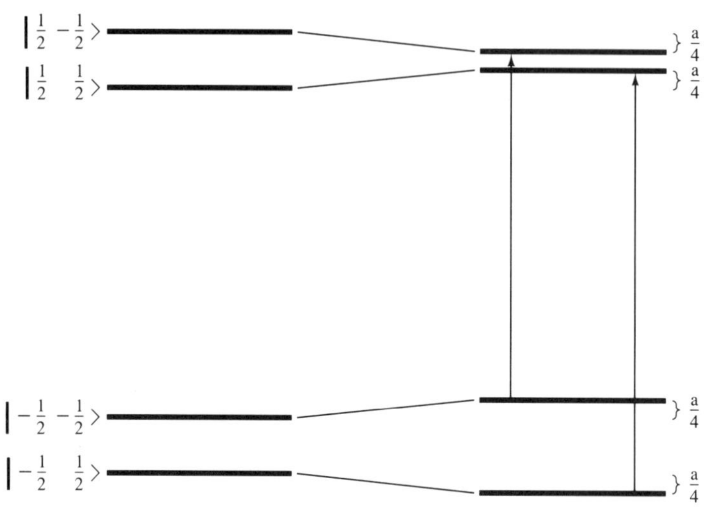

FIGURE 7.9 High-field energy levels for a system with electron spin S = 1/2 and with a magnetic nucleus with I = 1/2. The levels on the left are the zero-order energies, and the levels on the right include the first-order corrections due to electron spin-nuclear spin interaction. The two vertical arrows show the allowed transitions that would be observed in an ESR experiment. Not shown are the allowed transitions for which $M_S = 0$ since these are the NMR transitions and occur in a very different energy regime.

$$\hat{H}^{(1)} = a \vec{I} \cdot \vec{S} / \hbar^2 \qquad (7\text{-}20b)$$

The zero-order wavefunctions, labelled by the quantum numbers $m_S$ and $m_I$, are eigenfunctions of $\hat{I}_z$ and $\hat{S}_z$ (and of $\hat{I}^2$ and $\hat{S}^2$), and the associated energy values, represented in Fig. 7.9, are

| State $\mid m_S \; m_I >$ | $E^{(0)}_{m_S m_I}$ |
|---|---|
| 1/2 −1/2 | $g_e \mu_B H_z / 2 + g_H \mu_o H_z (1 - \sigma) / 2$ |
| 1/2 1/2 | $g_e \mu_B H_z / 2 - g_H \mu_o H_z (1 - \sigma) / 2$ |
| −1/2 −1/2 | $- g_e \mu_B H_z / 2 + g_H \mu_o H_z (1 - \sigma) / 2$ |
| −1/2 1/2 | $- g_e \mu_B H_z / 2 - g_H \mu_o H_z (1 - \sigma) / 2$ |

For each of these functions, the first-order energy corrections are

$$E^{(1)}_{m_S m_I} = < m_S\ m_I \mid \hat{H}^{(1)} \mid m_S\ m_I >$$

$$= \frac{a}{\hbar^2} < m_S\ m_I \mid \hat{I}_x \hat{S}_x + \hat{I}_y \hat{S}_Y + \hat{I}_z \hat{S}_z \mid m_S\ m_I >$$

$$= a\, m_S\, m_I \qquad (7\text{-}21)$$

A sketch of the first-order energy levels is given in Fig. 7.9. The high-field selection rules are $|\Delta m_S| = 1$ and $|\Delta m_I| = 0$. Thus, there will be two ESR transitions, and the difference in the two transition energies is simply the coupling constant a.

If there are several magnetic nuclei in a molecule, then the high-field energy level expression that follows from a first-order perturbative treatment of the general Hamiltonian of Eq. (7-19) is

$$E_{m_S m_{I_1} \cdots} = g_e \mu_B H_z m_S - \sum_i^{\text{nuclei}} g_i \mu_o H_z (1 - \sigma_i) m_{I_i}$$

$$+ \sum_{i>j}^{\text{nuclei}} J_{ij}\, m_{I_i}\, m_{I_j} + \sum_i^{\text{nuclei}} a_i\, m_S\, m_{I_i} \qquad (7\text{-}22)$$

Fig. 7.10 gives an example of energy levels that arise from Eq. (7-22) and the resulting form of the spectrum. Notice that there is a different multiplet pattern for the ESR transitions associated with an I = 1 nucleus than with an I = 1/2 nucleus. Indeed, the patterns of lines help identify the type of magnetic nuclei much as they do in NMR spectra, and this holds in the case of equivalent magnetic nuclei, too.

ESR transition energies yield values for each of the $a_i$ coupling constants in Eq. (7-22). The relative size of these constants is significant because the localization of the electron spin density must enter into determining the size of each $a_i$. In fact, ESR provides an ideal means for characterizing organic free radicals. If the measured coupling constants can be properly assigned to atoms in the radical, then one particularly large coupling constant suggests that there is an unpaired electron (the radical electron) that is largely localized at or near that atom. Of course, the

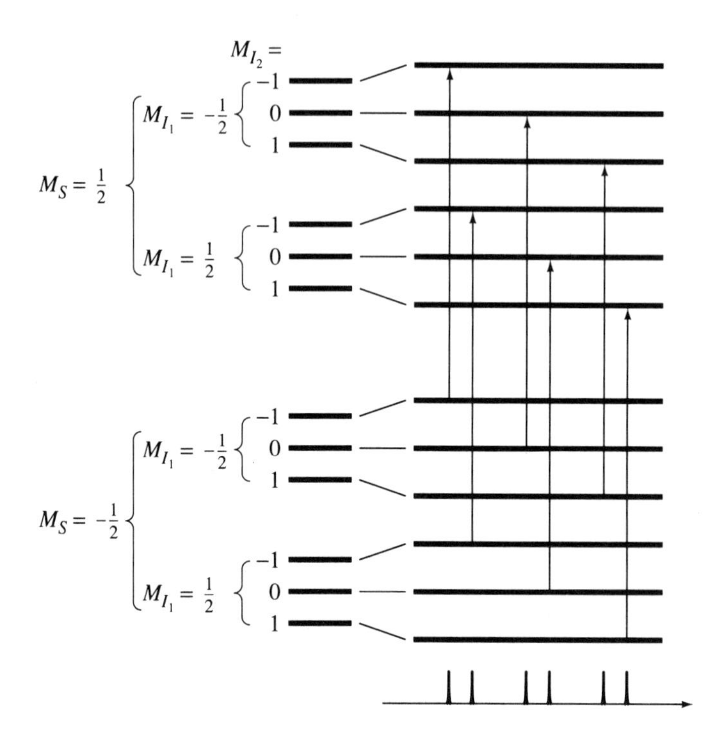


**FIGURE 7.10** High-field energy levels for a system with electron spin S = 1/2 and magnetic nuclei with I = 1/2 and I = 1. On the left are the levels at zero -rder where only the interaction with the external field is included. On the right are the energy levels according to Eq. (7-22) where the spin-spin interactions have been included by first-order perturbation theory. The nuclear spin-spin interaction has been exaggerated relative to typical values in order to show the effect. The vertical lines correspond to the allowed ESR transitions, and below them is a stick representation of the spectrum that corresponds to these transitions.

coupling constants might indicate that the unpaired electron is delocalized through the molecule, and this will have implications for the electronic structure of the species.

The low-field ESR problem is a more complicated quantum mechanical problem than the high-field case because first-order perturbation theory is not appropriate. A more suitable approach is to use linear variation theory with the Hamiltonian of Eq. (7-19) and with a basis set of independent spin functions, that is, product functions that are eigenfunctions of each of the z-component operators in Eq. (7-19). From the energy levels that result, the transition energies are obtained as differences in state energies. In practice, ESR spectra are sometimes analyzed by carrying out this process repeatedly for different choices of the coupling constants until a satisfactory agreement between measured and calculated transition energies is achieved. In this manner, the coupling constants may be extracted from the spectra even without the simplifying aspects of the high-field case.

We may understand some of the features of low-field ESR spectra by considering the low-field energy levels to be intermediate to two limiting cases, the high-field and the zero-field limits. In the absence of an external field, spins are fully coupled. So, for a problem with an electron spin, S, and a single magnetic nucleus with spin I, the rules of angular momentum coupling dictate the energy levels. The total angular momentum is the vector sum of the two spin vectors,

$$\vec{F} = \vec{S} + \vec{I} \tag{7-23}$$

and the associated quantum number F must have the values

$$F = |S - I|, \ldots, S + I$$

Then, as was done for spin-orbit interaction in the hydrogen atom, we may replace the spin-spin operator with the following,

$$\vec{I} \cdot \vec{S} = \frac{1}{2}\left(\hat{F}^2 - \hat{I}^2 - \hat{S}^2\right) \tag{7-24}$$

In the zero-field limit, the Hamiltonian reduces to

$$\hat{H} = \frac{a}{\hbar^2}\,\vec{I} \cdot \vec{S}$$

and so the eigenfunctions of the Hamiltonian are the functions that are simultaneously eigenfunctions of the three angular momentum operators $\hat{F}^2$, $\hat{S}^2$, and $\hat{I}^2$. Therefore, the eigenenergies must be

$$E_{F,S,I} = \frac{a}{2}\left[F(F+1) - I(I+1) - S(S+1)\right] \tag{7-25}$$

Each of these energy levels has a multiplicity (degeneracy) of 2F + 1.

For the specific situation of an electron spin S = 1/2 and a nuclear spin I = 1/2, F takes on the values of 0 and 1, and there are two energy levels. As the magnetic field is increased from zero to some high value, the energy levels will change smoothly from one limiting case to the high-field case where there are four energy levels. As the field is just turned on, the degeneracy in the levels of different F values is removed. For extremely weak fields, the magnetic field may be treated with low-order perturbation theory, and then the separation of levels will depend on the value of $m_F$, the quantum number that gives the net z-component of the total angular momentum. Thus, the F = 1 level will give rise to three levels ($m_F = -1, 0, 1$) that spread apart linearly with the field strength, at least initially.

## 7.4 MAGNETIC RESONANCE IMAGING

As described so far, magnetic resonance is a tool for probing individual molecules. However, there is also important information that develops as a bulk property, and this has become the basis for a powerful application, magnetic resonance imaging. Here, the general idea is to measure a signal that is dependent on the spatial distribution of a species that might otherwise be studied with conventional magnetic resonance spectroscopy. The mathematical task of imaging is to map the signal into a graphical respresentation, an image.

To illustrate the idea an imaging experiment, consider two long sample tubes filled with ethanol and held parallel in a larger diameter tube that is otherwise filled with benzene. This assembly is placed in an NMR spectrometer modified so that instead of the static magnetic field being uniform across the sample the field varies in strength from one side to the other. As a consequence of the field variation the ethanol NMR transition signals will vary with the orientation of two small tubes with respect to the field: If they are aligned with the applied field, one sample experiences a stronger field than the other, and so there will be signals at two distinct frequencies. If they are rotated away from this arrangement, the transition peaks will coalesce as the fields experienced by the samples

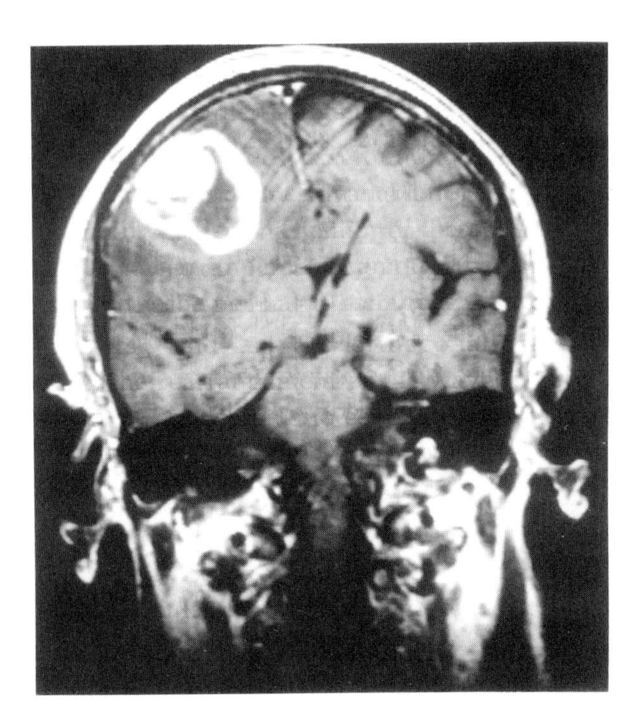

**FIGURE 7.11** An MRI image of an axial section of a human brain. The bright region in the right cerebral hemisphere is an indication of a tumor. The MRI used to obtain these images operates with a 1.5 Tesla magnet, and the image is based upon signals collected for a region that was 5 mm thick. (Photograph courtesy of the Department of Radiology, University of Iowa.)

in the two small tubes approach the same strength (i.e., at $90^\circ$ from the original orientation). In this way, a spatially varying field strength encodes spatial information into the NMR spectrum.

There is a bulk property of a sample that corresponds to the magnetic moment of an inidividual molecule called the *nuclear magnetization*, $\vec{M} = (M_x, M_y, M_z)$. It is the sum of the individual magnetic moment vectors per unit volume. The magnetization can be measured and its evolution in time can be followed. The characteristic times associated with the time evolution of the magnetization also encode spatial information if the field is non-uniform. There are a number of sophisticated means for measuring these times that involve a pulse sequence in the applied radiation.

Magnetic resonance imaging (MRI) translates spatially encoded NMR signals into images such as that in Fig. 7.11. The technology of MRI has advanced to the point that it is perhaps the non-destructive technique with the greatest resolution in biological systems.

## Exercises

1. Assuming all nuclei are in their ground state, how many different nuclear spin states will exist for the following isotopes of water: $H_2O$, $D_2O$, and $D_2{}^{17}O$?
2. Neglecting spin-spin coupling in the example of diimide, HNNH, of the 84 transitions, how many correspond to nitrogen spin flips, and how many correspond to hydrogen spin flips?
3. Neglecting spin-spin coupling, develop an energy level diagram for the nuclear spin states of an isotope of acetylene, $DC^{13}CH$, in a magnetic field. Assume that $g_D > g_C > g_H$. Draw vertical arrows to indicate the allowed NMR transitions.
4. Show that $< m_{I_1} m_{I_2} \mid \hat{I}_{x_1} \hat{I}_{x_2} + \hat{I}_{y_1} \hat{I}_{y_2} \mid m_{I_1} m_{I_2} >$ equals zero. (First, using the definitions of angular momentum raising and lowering operators, express the operator in the integral in terms of the raising and lowering operators.)
5. Develop a stick representation of the high resolution NMR spectra predicted by first-order treatment of spin-spin interaction of the following molecules.

   *a.* $H_2C^{17}O$ *b.* $H_2C{=}^{13}CH_2$
6. What is the intensity pattern for a proton NMR transition split by interaction with two nearby equivalent $^{14}N$ nuclei?
7. The NMR intensity/splitting pattern for a single magnetic nucleus is 1:1 with one nearby proton; it is 1:2:1 if there are two nearby equivalent protons, and 1:3:3:1 if there are three nearby equivalent protons. What would these patterns become if the protons were replaced by deuterons?

8. If there were four equivalent spin-1/2 nuclei in a molecule and a nearby proton that had a spin-spin interaction with these nuclei, what would be the intensity pattern for the NMR spectrum of that proton?
9. Predict the form of the $^{13}C$ and the proton NMR spectra of $H{-}^{13}C{\equiv}^{13}C{-}^{13}C{\equiv}^{13}C{-}H$ on the basis of a first-order treatment of spin-spin coupling.
10. Predict the high-field ESR spectrum of $H^{13}CO\bullet$ and of $H^{13}C^{13}C\bullet$.
11. Predict the high-field ESR spectrum of the radical $H_3C\text{-}CH_2\bullet$ assuming (a) that the $a_i$ coupling constants for $CH_2$ protons are much greater than for the $CH_3$ group protons, and (b) vice versa. Next, assume that the coupling constants are similar in size, although slightly different. Given that this is intermediate between the first two cases, what is the likely form of the spectrum?
12. Predict the high-field ESR spectrum of $HOO\bullet$, $H^{17}OO\bullet$, and $H^{17}O^{17}O\bullet$.
13. Consider the proton NMR spectra of the acetylenic hydrogens that are embedded in two identical larger molecules (e.g., the spectra of the HC≡C– fragments) that are separated in space in by 1 cm. Make a plot of the separation between the two transition peaks for the two molecules under the condition of a magnetic field gradient along the line connecting the two molecular centers. (Assume a value for the average field strength and plot the separation in the absorption signals as a function of the size of the gradient of the field.)

## Bibliography


1. M. Karplus and R. N. Porter, *Atoms and Molecules* (Benjamin, Menlo Park, California, 1970). This introductory text provides some of the ideas for understanding the contributing elements of chemical shielding.
2. W. H. Flygare, *Molecular Structure and Dynamics* (Prentice-Hall, Englewood Cliffs, New Jersey, 1978). This text offers an intermediate level development of magnetic resonance.

3. F. A. Bovey, L. Jelinski, and P. A. Mirau, *Nuclear Magnetic Resonance Spectroscopy* (Academic, New York, 1988). This provides an up-to-date, thorough coverage of the field including magnetic resonance imaging.

# Appendix I: Matrix Algebra

Matrices are ordered arrays of constants, variables, functions, or almost anything. A vector in a Cartesian coordinate system is a very simple ordered array; it consists of three elements which are the x-, y-, and z-components of the vector. The order of the elements tells which is which. In the vector (a,b,c), the x-component is a and not b or c. The ordering is established whether one uses a column or a row arrangement of the elements.

$$\text{column: } \begin{pmatrix} a \\ b \\ c \end{pmatrix} \qquad\qquad \text{row: } (a,b,c)$$

Of course, there may be more than three elements in certain cases, and so it is convenient to subscript the elements with a number that gives the position in the row or column, e.g., $(a_1, a_2, a_3, \ldots, a_n)$. A single column of elements is termed a *column matrix*. A letter with an arrow is the designation used herein for a column matrix (or vector). Usually the same letter is used for the individual elements, but they are subscripted:

$$\vec{a} = \begin{pmatrix} a_1 \\ a_2 \\ a_3 \\ a_4 \\ \cdots \\ a_n \end{pmatrix} \tag{I-1}$$

A row array of elements is a *row matrix*. Such a matrix will be designated in the same way as a column matrix, except that there will be a T (for transpose) as a superscript.

$$\vec{a}^T = \begin{pmatrix} a_1 & a_2 & a_3 & a_4 & \dots & a_n \end{pmatrix} \tag{I-2}$$

Other sets of elements may be ordered by both rows and columns, and these are *square* or *rectangular matrices*. These arrays are designated herein with boldface letters (usually capitals), and the elements are subscripted with a row-column index, that is, with two integers that give the row and column position in the array.

$$\mathbf{A} = \begin{pmatrix} A_{11} & A_{12} & A_{13} & \dots & A_{1n} \\ A_{21} & A_{22} & A_{23} & \dots & A_{2n} \\ A_{31} & A_{32} & A_{33} & \dots & A_{3n} \\ & & \dots & & \\ A_{m1} & A_{m2} & A_{m3} & \dots & A_{mn} \end{pmatrix} \tag{I-3}$$

This is an m-by-n matrix because there are m rows and n columns. Matrices can also be defined with three indices, or more.

Matrix addition is defined as the addition of corresponding elements of two matrices. The following two examples illustrate this definition.

$$\begin{pmatrix} 1 \\ 3 \\ 7 \\ 0 \end{pmatrix} + \begin{pmatrix} -1 \\ 0 \\ 10 \\ 2 \end{pmatrix} = \begin{pmatrix} 0 \\ 3 \\ 17 \\ 2 \end{pmatrix}$$

$$\begin{pmatrix} 0 & 3 & -i \\ 2 & -1 & 5 \end{pmatrix} + \begin{pmatrix} 0.5 & 1 & 1 \\ -2 & 0 & 5 \end{pmatrix} = \begin{pmatrix} 0.5 & 4 & 1-i \\ 0 & -1 & 10 \end{pmatrix}$$

The zero matrix is a matrix whose elements are all zero, and it is designated herein as **0** (bold zero).

Multiplication of a matrix by a scalar (a single value) is defined as the multiplication of every element in the matrix by that scalar. That is,

$$c\,\mathbf{A} = c\begin{pmatrix} A_{11} & A_{12} & A_{13} \\ A_{21} & A_{22} & A_{23} \\ A_{31} & A_{32} & A_{33} \end{pmatrix} = \begin{pmatrix} cA_{11} & cA_{12} & cA_{13} \\ cA_{21} & cA_{22} & cA_{23} \\ cA_{31} & cA_{32} & cA_{33} \end{pmatrix}$$

Multiplication of a matrix by another matrix goes by the "row-into-column" procedure. If **C** = **A B** , then the elements of **C** are given by

$$C_{ij} = \sum_{k=1}^{n} A_{ik} B_{kj} \tag{I-4}$$

where n is the number of columns of **A** *and* the number of rows of **B**. Recall that a vector dot product is,

$$(a_1 \ a_2 \ a_3 \ \ldots \ a_n) \begin{pmatrix} b_1 \\ b_2 \\ b_3 \\ \ldots \\ b_n \end{pmatrix} = a_1b_1 + a_2b_2 + a_3b_3 + \ldots + a_nb_n$$

Thus, Eq. (I-4) says that the i-j element of **C** is obtained by a vector dot product of the $i^{th}$ row of **A** and the $j^{th}$ column of **B**:

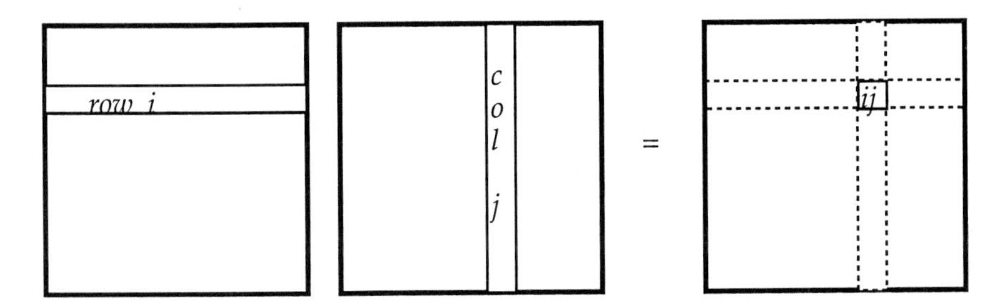


There are several features of matrix multiplication that may not be expected. First, it is not commutative. In general **A B** ≠ **B A**. Second, it is possible for the product of two matrices to be the zero matrix even if both have nonzero elements.

The identity for multiplication of square matrices is a matrix designated here as **1** (bold one). It is a matrix whose elements are all zero except for elements along the diagonal; these are one.

$$\mathbf{1} = \begin{pmatrix} 1 & 0 & 0 & \ldots \\ 0 & 1 & 0 & \ldots \\ 0 & 0 & 1 & \ldots \\ & \ldots & & \end{pmatrix} \tag{I-5}$$

It is easy to show for some arbitrary matrix **A**, that **A 1** = **1 A** = **A**.

The *transpose of a matrix* is formed by interchanging rows with columns. A transpose is designated here with a superscript T.

$$\mathbf{A} = \begin{pmatrix} 1 & 2 & 3 & 4 \\ 5 & 6 & 7 & 8 \\ 9 & 10 & 11 & 12 \\ 13 & 14 & 15 & 16 \end{pmatrix} \Rightarrow \mathbf{A}^T = \begin{pmatrix} 1 & 5 & 9 & 13 \\ 2 & 6 & 10 & 14 \\ 3 & 7 & 11 & 15 \\ 4 & 8 & 12 & 16 \end{pmatrix}$$

The complex conjugate of a matrix is the matrix of complex conjugates of the elements of the matrix.

$$\mathbf{B} = \begin{pmatrix} b_{11} & b_{12} & b_{13} \\ b_{21} & b_{22} & b_{23} \\ b_{31} & b_{32} & b_{33} \end{pmatrix} \Rightarrow \mathbf{B}^* = \begin{pmatrix} b^*_{11} & b^*_{12} & b^*_{13} \\ b^*_{21} & b^*_{22} & b^*_{23} \\ b^*_{31} & b^*_{32} & b^*_{33} \end{pmatrix}$$

The complex conjugate transpose of a matrix is called the adjoint and is designated with a superscript †. Thus, $\mathbf{A}^\dagger = \mathbf{A}^{T*}$.

The inverse of some matrix $\mathbf{C}$ is $\mathbf{C}^{-1}$ if the following is true:

$$\mathbf{C}\mathbf{C}^{-1} = \mathbf{C}^{-1}\mathbf{C} = \mathbf{1}$$

$\mathbf{C}$ and $\mathbf{C}^{-1}$ are inverses of each other.

A *determinant* is a scalar quantity that is expressed in terms of $n^2$ elements where n is the order of the determinant. The elements are arranged in n rows and n columns and placed between vertical bars. Its appearance is similar to a square matrix, but it is something entirely different since it represents a single value. That value is obtained from the $n^2$ elements. Carrying out the evaluation involves using the *minor* of an element of a determinant. To see how this is done, let $d_{ij}$ be an arbitrary element of the determinant $\Phi$:

$$\Phi = \begin{vmatrix} d_{11} & d_{12} & d_{13} & \cdots & d_{1n} \\ d_{21} & d_{22} & d_{23} & \cdots & d_{2n} \\ \cdot & \cdot & \cdot & \cdots & \cdot \\ d_{n1} & d_{n2} & d_{n3} & \cdots & d_{nn} \end{vmatrix}$$

The minor of the element $d_{ij}$ is the determinant that remains after deleting the $i^{th}$ row and $j^{th}$ column from $\Phi$. We shall designate that determinant $D^{ij}$. The value of the determinant $\Phi$ is the sum of the products of elements from any row, or from any column, and their minors with a particular factor of –1:

$$\Phi = \sum_{i=1}^{n} (-1)^{i+j} d_{ji} D^{ji} \quad \text{for } j = 1, 2, \ldots, n$$

$$\Phi = \sum_{i=1}^{n} (-1)^{i+j} d_{ij} D^{ij} \quad \text{for } j = 1, 2, \ldots, n \qquad \text{(I-6)}$$

Of course, finding values for each of the $D^{ij}$ determinants also requires using one of these equations. The process reaches a conclusion, though, when the minor has one only one element; the value of that determinant (the minor) is that element.

To illustrate the evaluation process, notice that a second-order determinant is evaluated from Eq. (I-6) in the following way:

$$\begin{vmatrix} e_{11} & e_{12} \\ e_{21} & e_{22} \end{vmatrix} = (-1)^{1+1} e_{11} E^{11} + (-1)^{1+2} e_{12} E^{12}$$

$$= e_{11} e_{22} - e_{12} e_{21}$$

This is helpful because applying Eq. (I-6) to a third-order determinant (one with three rows and three columns) would give a sum with minors that are second-order determinants; this result can be used in working out the value for a third-order determinant. A fourth-order determinant could be evaluated with minors that are third order, and in such a stepwise fashion, a determinant of any order can be evaluated with Eq. (I-6). When carried out fully, the value of a determinant of order n will be a sum of n! products of its elements.

There is a linear algebra equivalent of the differential eigenvalue equation. Instead of a differential operator acting on a function, a square matrix multiplies a column vector, and this is equal to a constant, the eigenvalue, times the vector:

$$\mathbf{A}\vec{c} = e\vec{c} \qquad \text{(I-7)}$$

**A** is a square matrix, $\vec{c}$ is the *eigenvector*, and e is the eigenvalue. For an N-by-N matrix, there are N different eigenvalues and N different eigenvectors. That is, there are N different solutions to Eq. (I-7). Thus, the equation is better written with subscripts that distinguish the different solutions:

$$\mathbf{A}\,\vec{c}_i = e_i\,\vec{c}_i \tag{I-8}$$

Because i = 1, ..., N, Eq. (I-8) does represent the N different equations. However, from the rules for matrix multiplication, it is easy to realize that these equations may be collectively represented by one matrix equation. This is accomplished by collecting the column eigenvectors into one N-by-N matrix.

$$\mathbf{A}\begin{pmatrix}\vec{c}_1 & \vec{c}_2 & \vec{c}_3 & \cdots & \vec{c}_N\end{pmatrix} = \begin{pmatrix}\vec{c}_1 & \vec{c}_2 & \vec{c}_3 & \cdots & \vec{c}_N\end{pmatrix}\begin{pmatrix} e_1 & 0 & 0 & \cdots & 0 \\ 0 & e_2 & 0 & \cdots & 0 \\ 0 & 0 & e_3 & \cdots & 0 \\ & & & \cdots & \\ 0 & 0 & 0 & \cdots & e_N \end{pmatrix} \tag{I-9}$$

With the different column vectors set next to each other, we obtain a square array, which will be designated **C**. The rightmost matrix in Eq. (I-9), which will be called **E**, is a *diagonal matrix* since its only nonzero elements are along the diagonal. Because of its diagonal form, matrix multiplication of **E** by **C** just leads to each column of **C** scaled by a corresponding diagonal element of **E**. From this, we have the form of the general matrix eigenvalue equation,

$$\mathbf{A\,C} = \mathbf{C\,E} \quad \text{where } \mathbf{E} \text{ is diagonal} \tag{I-10}$$

This equation is solved for some specific matrix **A**, and the solutions are the eigenvectors arranged as columns in **C**, plus the associated eigenvalues arranged on the diagonal of **E**.

It is quite often the case that a matrix whose eigenvalues are sought is real and symmetric, and we will only consider such matrices here. If the matrix **C** in Eq. (I-10) were known, and if its inverse could be found, then multiplication of Eq. (I-10) by the inverse would lead to,

$$\mathbf{C}^{-1}\mathbf{A\,C} = \mathbf{C}^{-1}\mathbf{C\,E} = \mathbf{E} \tag{I-11}$$

The process of multiplying some matrix on the right by a particular matrix, and multiplying on the left by the inverse of that matrix, is called a *transformation*.

The general matrix eigenequation, expressed in the form of Eq. (I-11), can now be thought of differently: We seek a matrix **C** that transforms the matrix **A** in a very particular way such that the transformed matrix is of diagonal form. The process to accomplish this is referred to simply as *diagonalization* of **A**. If we seek the eigenvalues and eigenvectors of some matrix, then we seek to diagonalize it.

An important category of transformation matrices is that of *unitary matrices*. These are employed for diagonalizing real, symmetric matrices, and we will use **U** to designate a unitary matrix in the following discussion. A unitary matrix is one whose inverse is the same as or equal to its transpose. Or, if the elements of **U** are complex, then unitarity means the inverse is equal to the adjoint: $\mathbf{U}^{-1} = \mathbf{U}^{\dagger}$. The property of unitarity means that a transformation can be written with either the transpose (or adjoint) or with the inverse.

$$\mathbf{U}^{-1}\mathbf{A}\,\mathbf{U} = \mathbf{U}^{T}\mathbf{A}\,\mathbf{U}$$

Another property of a unitary matrix is that the value of the determinant constructed from its elements is equal to one; that is, $\det(\mathbf{U}) = 1$.

There are many ways of finding unitary matrices to diagonalize real, symmetric matrices. A very useful recipe for 2-by-2 matrices uses sines and cosines. Starting with **A** as a matrix we seek to diagonalize

$$\mathbf{A} = \begin{pmatrix} A_{11} & A_{12} \\ A_{21} & A_{22} \end{pmatrix}$$

an angle θ is determined from values in **A**.

$$\tan 2\theta = \frac{2\,A_{12}}{A_{22} - A_{11}} \tag{I-12}$$

The transformation matrix is,

$$\mathbf{U} = \begin{pmatrix} \cos\theta & \sin\theta \\ -\sin\theta & \cos\theta \end{pmatrix} \tag{I-13}$$

The eigenvalues are

$$\mathbf{U}^{T}\mathbf{A}\,\mathbf{U} = \begin{pmatrix} e_1 & 0 \\ 0 & e_2 \end{pmatrix} \tag{I-14}$$

Diagonalization by 2-by-2 rotations is one means of diagonalization in larger problems. The idea is to extract a 2-by-2 from a large matrix and use the recipe above. To see what this involves, we must recall special features of matrix multiplication. Consider an 8-by-8 matrix, **A**. Let us extract the following 2-by-2 from it.

$$\begin{pmatrix} A_{33} & A_{36} \\ A_{63} & A_{66} \end{pmatrix}$$

And let the matrix that diagonalizes this be the following.

$$\mathbf{U} = \begin{pmatrix} a & b \\ c & d \end{pmatrix}$$

In order to apply the transformation given by **U** to the entire **A** matrix, it must be superposed on the unit matrix. This means that we replace the 3-3, 3-6, 6-3 and 6-6 elements of an 8-by-8 unit matrix (i.e., **1**) by a, b, c and d, respectively, from **U**. These replaced elements are in the very same positions as the elements extracted from **A** to make up the 2-by-2 for diagonalization. As illustrated below, this matrix multiplying an 8-by-8 square matrix will alter the values of only the elements in rows 3 and 6.

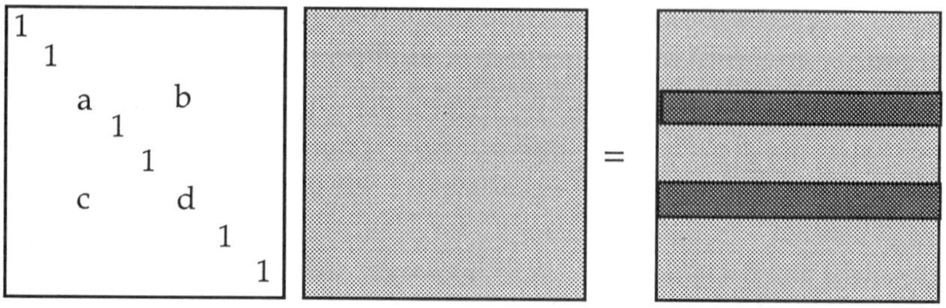

$$\begin{bmatrix} 1 &&&&&&& \\ & 1 &&&&&& \\ && a &&& b && \\ &&& 1 &&&& \\ &&&& 1 &&& \\ && c &&& d && \\ &&&&&& 1 & \\ &&&&&&& 1 \end{bmatrix} \; [\;\text{shaded}\;] \; = \; [\;\text{shaded}\;]$$

The superposed unit matrix is shown on the left (only nonzero values) multiplying another matrix (shaded). The only elements that are different after the matrix multiplication are those in the rows with a, b, c, and d, and these are indicated by the darkened rows on the rightmost matrix. Were the multiplication done in the reverse order, the elements of the third and sixth columns would be the ones affected:

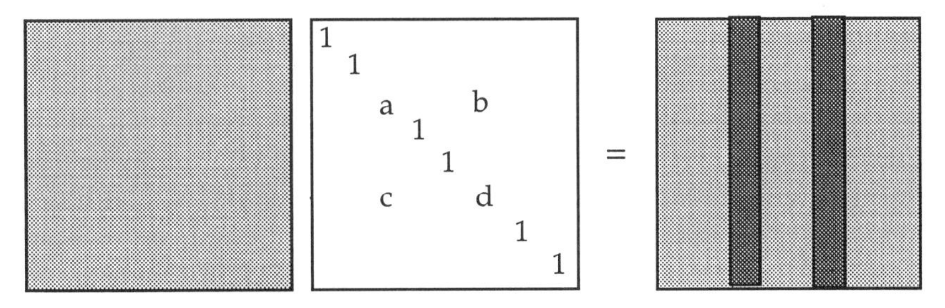


Thus, when the 2-by-2 **U** matrix is properly superposed on the unit matrix and used to transform the original **A** matrix, the result will be the following.

*Transformed* **A** *matrix:*

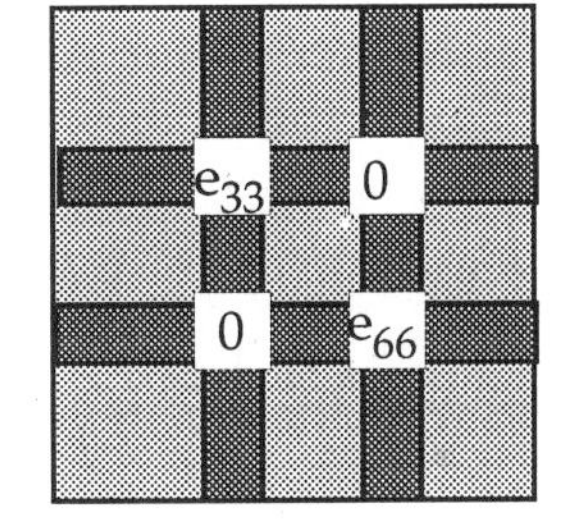


The darkened rows and columns indicate elements whose values are now different from what they were in the original **A** matrix. The eigenvalues of the extracted 2-by-2 part of the **A** matrix are now on the diagonal, and corresponding off-digaonal elements are zero, as indicated. This matrix is closer to being in diagonal form.

In the next step, another 2-by-2 is extracted from **A**, and the same thing is done. A complication is that in this second 2-by-2 rotation, the 3-6 and 6-3 elements that had become zero may change. That is, they may end up being different from the desired value of zero if the second 2-by-2 rotation involves the third or the sixth row and column. However, by continuing the process again and again, it is possible for all the off-diagonal elements to be as close to zero as one wishes. Typically, it is best to select the particular 2-by-2 for each step on the basis of the largest off-diagonal value present. This is a process which is easily coded for computers.

# Appendix II: Curve Fitting

Usually a mathematical relationship between an observable and a variable does not emerge directly from an experiment. More likely the experiment yields a set of data points, (x,y), where x is the variable, a controllable quantity set by the experimenter, and y is the observed or measured quantity. The problem of finding a functional relation between the observable and the variable, that is finding y(x), is the problem of curve fitting.

We must realize at the outset that the data, the set of (x,y) points, are not likely to be precise. There may be measurement uncertainty that limits the precision of the values of x and y. We must realize that the number of data points is finite, and so that means a minor feature (a bump or wiggle) that occurs for the true y(x) between two data points may not be revealed by the data. We won't know that it is there, and likewise we cannot know that it is not there. Furthermore, the data points themselves do not identify the actual functional form. For instance, for a set of three data points, there are many different functions that could pass through all three, perhaps sin(x), or a parabola, or an exponential. So, the task of curve fitting is nothing more than finding some analytical function that represents fairly well the set of data points.

One of the most standard approaches for curve fitting is the *least squares method*. The idea behind this method is that we choose a function with a number of embedded parameters. The function will tend to "hit and miss" the data points, but we may adjust the embedded parameters to make it "hit" more than it "misses." In the least squares method, the squares of the deviations of the data points from the curve given by our chosen function are summed. This value provides a measure of how well the function is representing the data. It is this value that we seek to minimize through adjustment of the embedded parameters. In the end we will have found the parameter choices that give the least squares error.

Let us assume that there are N data points: $(x_1, y_1)$, $(x_2, y_2)$, ..., $(x_N, y_N)$. We seek to find a function of x that closely represents the data. Let f(x) be a guess of the function. It should be a function that has one or more adjustable parameters which constitute the set $\{c_i\}$. For any particular data point, the error in the fit is the difference between y and f(x) for that point. If we were to sum the errors at each of the points, we would have a poor measure of the quality of the guess function. There could be one point where f(x) is much greater than the y value, and another point where f(x) is as much less than the y value. The sum of these two errors would be zero, and that would not reflect the fact that f(x) is not representing these two points well. For this reason, it is the square of the errors at each point that are important. The total of the squared errors shall be designated E, and so,

$$E = \sum_{i=1}^{N} (f(x_i) - y_i)^2 \qquad \text{(II-1)}$$

Least squares fitting means finding the choice of the parameters such that E is minimized. Clearly, E is a function of the parameters; as they are changed, E changes. When a function is at a minimum value, its first derivative (or derivatives) are zero. Recall that for a single variable function, the first derivative gives the slope of a line tangent to the curve of the function. At a minimum in the function, the curve is uphill in either direction, and so the tangent line must be flat. The slope is zero and the first derivative is zero.

Differentiation of Eq. (II-1) with respect to one of the adjustable parameters yields,

$$\frac{\partial E}{\partial c_j} = \sum_{i=1}^{N} 2\left( f(x_i) - y_i \right) \frac{\partial f(x_i)}{\partial c_j} \qquad \text{(II-2)}$$

At the choice of the $c_j$ parameters that gives the minimum value of E [i.e., the best f(x)], all the derivatives represented by Eq. (II-2) must be zero. Setting each equal to zero establishes a set of coupled equations that determine the optimum parameter choices. For large data sets, the solution of the coupled equations is conveniently done with computer programs.

If we do not know the functional form that some set of data should obey, it is always possible to try a power series.

$$f(x) = c_0 + c_1 x + c_2 x^2 + c_3 x^3 + \dots \tag{II-3}$$

This must be truncated at some point, and so we may fit the data to a quadratic function if we truncate after the $c_2$ term, or to a cubic function, or a quartic function, and so on. For this type of function, the adjustable parameters, $c_0$, $c_1$, $c_2$, and so on, enter the function linearly. Thus, the first derivative expressions are particularly simple.

$$\frac{\partial f}{\partial c_0} = 1$$

$$\frac{\partial f}{\partial c_1} = x$$

$$\frac{\partial f}{\partial c_2} = x^2 \qquad \text{and so on}$$

As an example, consider a quadratic fit of some set of data. Using the above first derivative expressions in Eq. (II-2) gives

$$\frac{\partial E}{\partial c_0} = \sum 2\,(f(x_i) - y_i) = 2\sum (c_0 + c_1 x_i + c_2 x_i^2 - y_i) \tag{II-4}$$

$$\frac{\partial E}{\partial c_1} = \sum 2(f(x_i) - y_i)\, x_i = 2\sum (c_0 x_i + c_1 x_i^2 + c_2 x_i^3 - x_i y_i) \tag{II-5}$$

$$\frac{\partial E}{\partial c_2} = \sum 2\,(f(x_i) - y_i)\, x_i^2 = 2\sum (c_0 x_i^2 + c_1 x_i^3 + c_2 x_i^4 - x_i^2 y_i) \tag{II-6}$$

Notice that the summations in these last three equations can be separated into summations over data values.

$$\sum_{i=1}^{N} (c_0 + c_1 x_i + c_2 x_i^2 - y_i) = N c_0 + c_1 \sum_{i=1}^{N} x_i + c_2 \sum_{i=1}^{N} x_i^2 - \sum_{i=1}^{N} y_i$$

It is now convenient to define total values of certain x-y products as,

$$\overline{x^n y^m} \equiv \sum_{i=1}^{N} x_i^n y_i^m \qquad \text{(II-7)}$$

These are simple values or numbers that can be obtained from the set of data points. Thus, Eq. (II-4) becomes

$$\frac{\partial E}{\partial c_0} = 2\left\{ N c_0 + c_1 \overline{x} + c_2 \overline{x^2} - \overline{y} \right\}$$

Finally, setting each of the three derivatives equal to zero, we arrive at three coupled equations.

$$N c_0 + c_1 \overline{x} + c_2 \overline{x^2} - \overline{y} = 0$$

$$c_0 \overline{x} + c_1 \overline{x^2} + c_2 \overline{x^3} - \overline{xy} = 0$$

$$c_0 \overline{x^2} + c_1 \overline{x^3} + c_2 \overline{x^4} - \overline{x^2 y} = 0$$

These are three linear equations in $c_0$, $c_1$ and $c_2$, and their solution gives the optimum parameter values.

# Appendix III: Table of Integrals

1. General integral relations

$$\int c\, f(x)\, dx = c \int f(x)\, dx$$

$$\int \left( f_1(x) + f_2(x) + \ldots \right) dx = \int f_1(x)\, dx + \int f_2(x)\, dx + \ldots$$

$$\int f(x)\, g(y)\, dx\, dy = \int f(x)\, dx \int g(y)\, dy$$

$$\int \left( f_1(x) + f_2(x) \right) \left( g_1(x) + g_2(x) \right) dx = \int f_1(x)\, g_1(x)\, dx$$

$$+ \int f_1(x)\, g_2(x)\, dx + \int f_2(x)\, g_1(x)\, dx + \int f_2(x)\, g_2(x)\, dx$$

2. Integrals over Gaussian functions ( $c > 0$ )

$$\int_0^\infty e^{-cx^2}\, dx = \frac{1}{2} \sqrt{\frac{\pi}{c}}$$

$$\int_0^\infty x\, e^{-cx^2}\, dx = \frac{1}{2c}$$

$$\int_0^\infty x^2 e^{-cx^2}\,dx = \frac{1}{4}\sqrt{\frac{\pi}{c^3}}$$

$$\int_0^\infty x^{2n+1} e^{-cx^2}\,dx = \frac{n!}{2\,c^{n+1}}$$

$$\int_0^\infty x^{2n} e^{-cx^2}\,dx = \frac{(2n-1)\,(2n-3)\ldots(3)\,(1)}{2^{n+1}}\sqrt{\frac{\pi}{c^{2n+1}}}$$

$$\int_{-\infty}^0 x^n e^{-cx^2}\,dx = (-1)^n \int_0^\infty x^n e^{-cx^2}\,dx$$

2. Integrals of exp(x) $(c > 0)$

$$\int x^n e^{cx}\,dx = e^{cx}\sum_{j=0}^{n}(-1)^j \frac{n!}{(n-j)!}\frac{x^{n-j}}{c^{j+1}}$$

$$\int_0^\infty x^n e^{-cx}\,dx = \frac{n!}{c^{n+1}} \qquad (\text{for } n = 1, 2, 3, \ldots)$$

3. Integrals over trigonometric functions

$$\int (\sin^2 cx)\,dx = \frac{x}{2} - \frac{\sin 2cx}{4c}$$

$$\int (\sin^3 cx)\,dx = -\frac{(\cos cx)\,(\sin^2 cx + 2)}{3c}$$

$$\int (\sin^n cx)\,dx = -\frac{(\sin^{n-1} cx)\,(\cos cx)}{nc} + \frac{n-1}{n}\int(\sin^{n-2} cx)\,dx$$

$$\int c\,x^n \sin(cx)\,dx = -x^n\cos(cx) + n\int x^{n-1}\cos(cx)\,dx$$

$$\int c\, x^n \cos(cx)\, dx = x^n \sin(cx) - n \int x^{n-1} \sin(cx)\, dx$$

$$\int x\, (\sin^2 cx)\, dx = \frac{x^2}{4} - \frac{x\,(\sin 2cx)}{4c} - \frac{\cos 2cx}{8c^2}$$

$$\int x^2\, (\sin^2 cx)\, dx = \frac{x^3}{6} - \left(\frac{x^2}{4c} - \frac{1}{8c^3}\right)(\sin 2cx) - \frac{x\,(\cos 2cx)}{4c^2}$$

# Appendix IV: Table of Atomic Masses and Nuclear Spins

This is a table of the isotopic masses and electron orbital occupancies of most of the non-transition elements of the periodic table. The values of the masses (in amu), the percent natural abundance of each isotope, and the nuclear spins are from the "Table of Nuclides" in *Nuclear and Radiochemistry*, 2nd ed., G. Friedlander, J. W. Kennedy, and J. M. Miller, (John Wiley and Sons, New York, 1964). Unstable isotopes and isotopes with an abundance of less than 0.1% have not all been included. Nuclear g factors [see Eq. (7-10)] are from the table "Nuclear Spins, Moments, and Magnetic Resonance Frequencies" in *Handbook of Chemistry and Physics*, 63rd ed., (CRC Press, Boca Raton, Florida, 1982).

| Element Number | Symbol | Orbital Occupancy | Isotope Mass | Nuclear Spin | $g_N$ | Percent Abund. |
|---|---|---|---|---|---|---|
| 1 | H | $1s^1$ | 1.007825 | 1/2 | 2.79268 | 99.985 |
| | | | 2.014102 | 1 | 0.85739 | 0.015 |
| | | | 3.016049 | 1/2 | 2.97877 | |
| 2 | He | $1s^2$ | 4.002604 | 0 | | 100 |
| 3 | Li | [He] $2s^1$ | 6.015126 | 1 | 0.8219 | 7.42 |
| | | | 7.016005 | 3/2 | 3.2560 | 92.58 |

| Element Number | Symbol | Orbital Occupancy | Isotope Mass | Nuclear Spin | $g_N$ | Percent Abund. |
|---|---|---|---|---|---|---|
| 4 | Be | [He] $2s^2$ | 9.012186 | 3/2 | –1.1774 | 100 |
| 5 | B | [He] $2s^2\,2p^1$ | 10.012939 | 3 | 1.8007 | 19.6 |
| | | | 11.009305 | 3/2 | 2.6880 | 80.4 |
| 6 | C | [He] $2s^2\,2p^2$ | 12.000000 | 0 | | 98.89 |
| | | | 13.003354 | 1/2 | 0.7022 | 1.11 |
| 7 | N | [He] $2s^2\,2p^3$ | 14.003074 | 1 | 0.4035 | 99.63 |
| | | | 15.000108 | 1/2 | –0.2830 | 0.37 |
| 8 | O | [He] $2s^2\,2p^4$ | 15.994915 | 0 | | 99.759 |
| | | | 16.999133 | 5/2 | –1.8930 | 0.037 |
| | | | 17.999160 | 0 | | 0.204 |
| 9 | F | [He] $2s^2\,2p^5$ | 18.998405 | 1/2 | 2.6273 | 100 |
| 10 | Ne | [He] $2s^2\,2p^6$ | 19.992440 | 0 | | 90.92 |
| | | | 20.993849 | 3/2 | –0.6614 | 0.26 |
| | | | 21.991385 | 0 | | 8.82 |
| 11 | Na | [Ne] $3s^1$ | 22.989773 | 3/2 | 2.2161 | 100 |
| 12 | Mg | [Ne] $3s^2$ | 23.985045 | 0 | | 78.70 |
| | | | 24.985840 | 5/2 | –0.8545 | 10.13 |
| | | | 25.982591 | 0 | | 11.17 |
| 13 | Al | [Ne] $3s^2\,3p^1$ | 26.981535 | 5/2 | 3.6385 | 100 |
| 14 | Si | [Ne] $3s^2\,3p^2$ | 27.976927 | 0 | | 92.21 |
| | | | 28.976491 | 1/2 | –0.5548 | 4.70 |
| | | | 29.973761 | 0 | | 3.09 |
| 15 | P | [Ne] $3s^2\,3p^3$ | 30.973763 | 1/2 | 1.1305 | 100 |
| 16 | S | [Ne] $3s^2\,3p^4$ | 31.972074 | 0 | | 95.00 |
| | | | 32.971460 | 3/2 | 0.6426 | 0.76 |
| | | | 33.967864 | 0 | | 4.22 |
| | | | 35.967090 | 0 | | 0.014 |

| Element Number | Symbol | Orbital Occupancy | Isotope Mass | Nuclear Spin | $g_N$ | Percent Abund. |
|---|---|---|---|---|---|---|
| 17 | Cl | [Ne] $3s^2\,3p^5$ | 34.968854 | 3/2 | 0.8209 | 75.53 |
| | | | 36.965896 | 3/2 | 0.6833 | 24.47 |
| 18 | Ar | [Ne] $3s^2\,3p^6$ | 35.967548 | 0 | | 0.337 |
| | | | 39.962384 | 0 | | 99.600 |
| 19 | K | [Ar] $4s^1$ | 38.963714 | 3/2 | 0.3910 | 93.10 |
| | | | 40.961835 | 3/2 | 0.2146 | 6.88 |
| 20 | Ca | [Ar] $4s^2$ | 39.962589 | 0 | | 96.97 |
| | | | 41.958628 | | | 0.64 |
| | | | 42.958780 | 7/2 | −1.3153 | 0.15 |
| | | | 43.955490 | | | 2.06 |
| 21 to 30 | Sc to Zn | [Ca] $3d^{1\text{-}10}$ | | | | |
| 31 | Ga | [Zn] $4p^1$ | 68.92568 | 3/2 | 2.011 | 60.4 |
| | | | 70.92484 | 3/2 | 2.5549 | 39.6 |
| 32 | Ge | [Zn] $4p^2$ | 69.92428 | 0 | | 20.52 |
| | | | 71.92174 | 0 | | 27.43 |
| | | | 72.9234 | 9/2 | −0.8768 | 7.76 |
| | | | 73.92115 | 0 | | 36.54 |
| | | | 75.9214 | 0 | | 7.76 |
| 33 | As | [Zn] $4p^3$ | 74.92158 | 3/2 | 1.4349 | 100 |
| 34 | Se | [Zn] $4p^4$ | 75.91923 | 0 | | 9.02 |
| | | | 76.91993 | 1/2 | 0.5325 | 7.58 |
| | | | 77.91735 | 0 | | 23.52 |
| | | | 79.91651 | 0 | | 49.82 |
| | | | 81.9167 | 0 | | 9.19 |
| 35 | Br | [Zn] $4p^5$ | 78.91835 | 3/2 | 2.0990 | 50.54 |
| | | | 80.91634 | 3/2 | 2.2626 | 49.46 |
| 36 | Kr | [Zn] $4p^6$ | 79.91639 | | | 2.27 |
| | | | 81.91348 | 0 | | 11.56 |
| | | | 82.91413 | 9/2 | −0.9671 | 11.55 |
| | | | 83.91150 | 0 | | 56.90 |
| | | | 85.91062 | 0 | | 17.37 |

# Appendix V: Fundamental Constants and Units Conversion

*Systems of Units.* A system of units of measure for mechanical systems may be defined by specifying the unit of measurement for just three physical quantities. For instance, if we decide how mass, length, and time are to be measured, then there will exist a basis for measurement of velocity (i.e., unit length per unit time), momentum, acceleration, energy, and so on. We may specify units of measurement for three other physical quantities to define a system of units, but not more than three.

In the metric system, the three familiar basic units are the gram, the meter, and the second, and the symbols are g, m, and s. Prefixes are used to designate units that are smaller by powers of 10 or larger by powers of 10. The commonly encountered prefixes in chemical physics are the following.

| *A metric prefix of (symbol):* | *Means the prefixed unit is scaled by:* |
|---:|---|
| tera (T) | $10^{12}$ |
| giga (G) | $10^{9}$ |
| mega (M) | $10^{6}$ |
| kilo (k) | 1000 |
| centi (c) | 0.01 |
| milli (m) | 0.001 |
| micro (μ) | $10^{-6}$ |

| | |
|---|---|
| nano (n) | $10^{-9}$ |
| pico (p) | $10^{-12}$ |
| femto (f) | $10^{-15}$ |

So, 1 mg is 0.001 of a gram, and 1 kg is 1000 g.

There are two traditional selections of a basic unit for mass and for length that define units of other physical quantitities. The "mks" form uses meters, kilograms, and seconds. The mks energy unit, called the joule, must be the mass unit (kg), the length unit (m) to the second power, and the inverse of the second power of the time unit (s).

$$1\,\text{J} = 1\ \text{kg}\,(\text{m/s})^2$$

The "cgs" form uses centimeters, grams, and seconds. The cgs energy unit is the erg:

$$1\ \text{erg} = 1\ \text{g}\,(\text{cm/s})^2 = 10^{-3}\ \text{kg}\,(10^{-2}\ \text{m/s})^2 = 10^{-7}\ \text{J}$$

The conversion between mks and cgs for any other physical quantity is also obtained by going back to the relative sizes of the mass and length quantities .

*Fundamental Constants.* Among the "properties of nature," we may include the values of fundamental constants. Three are particularly important in the quantum mechanics of atoms and molecules. These are the speed of light (c), Planck's constant ($\hbar$), and the mass of an electron ($m_e$). The values of these constants in the cgs units are

$$c = 2.997925 \times 10^{10}\ \text{cm/sec}$$

$$\hbar = 1.05450 \times 10^{-27}\ \text{erg-sec} \quad (\text{also used: } h = 2\pi\hbar\,)$$

$$m_e = 9.1091 \times 10^{-28}\ \text{g}$$

*Frequencies.* Frequencies are associated with a regular oscillation in time. They are expressed in two ways, angular frequencies and linear frequencies. Angular frequencies are in radians per unit time, whereas linear frequencies are in cycles per unit time. Though there is no standard notation convention, it is frequently the case that $\omega$ is used for an angular frequency and $\nu$ is used for a frequency expressed in cycles per unit time. The relation between the two is simple since the number of radians covered in one cycle must be $2\pi$, the number of radians in a circle.

$$2\pi\nu = \omega$$

An oscillation in time (t) that followed a cosine function would be written as either cos($\omega$t) or cos(2$\pi\nu$t). That is usually the context that distinguishes angular from linear frequencies when the choice is not stated explicitly. The frequency of electromagnetic radiation is usually specified by the cycles per second, a unit that is called the hertz, and so the values are linear frequencies.

The energy, E, of a photon of radiation is directly proportional to the frequency, and the proportionality constant is Planck's constant:

$$E = \hbar\omega = h\nu$$

Thus, the frequency scale of cycles per second may be taken to be equivalent to an energy scale in ergs (via the factor h).

For historical reasons from the development of spectroscopy, another frequency scale has proved to convenient. In this case, the frequencies, $\nu$, are divided by a fundamental constant, the speed of light, c. Then, the frequencies are given in the inverse unit of length, $cm^{-1}$, and 1 $cm^{-1}$ is called a *wavenumber*. This amounts to using the relation between wavelength ($\lambda$) of radiation and frequency, $c = \lambda\nu$, so as to give the inverse wavelength instead of the frequency, the two being proportional:

$$\frac{1}{\lambda} = \frac{\nu}{c}$$

Wavenumbers serve as an energy scale just as well as frequencies do, and it is very common to give atomic and molecular energies in units of $cm^{-1}$.

$$\tilde{E}\,(cm^{-1}) = \frac{E(ergs)}{hc} = \frac{\nu}{c} = \frac{\omega}{2\pi c} \equiv \tilde{\omega}\,(cm^{-1})$$

The symbol $\tilde{E}$ as opposed to just E is sometimes used to indicate that the system of units for the energy is wavenumbers and not ergs. A special symbol, of course, is unnecessary because an energy expression must hold in any system of units.

*Electromagnetic Units and Constants.* From an atomic and molecular view, one of the most basic electromagnetic quantities is the charge of an electron. This fundamental quantity is usually designated e. The size of the charge of an electron in the International System of Units (SI) is $1.6021892 \times 10^{-19}$ coulombs (C). In the cgs-emu (centimeter, gram, second–electromagnetic unit) system, the value is $1.6021892 \times 10^{-20}$ emu, or 10 times the value in coulombs. (This means 1 emu = 10 C; an emu of charge is sometimes referred to as an abcoulomb.) The SI definition of a coulomb is that it is the charge that gives a 1-newton (N) force when placed 1 m from a like charge. Thus, the unit of charge can be expressed in terms of mass, length, and time units. In particular, 1 emu = 1 $cm^{1/2}\,g^{1/2}$.

The SI unit for electrical potential is the volt (V). A particle experiencing an electrical potential will have an interaction energy (in joules) that is equal to the product of the charge in SI units and the electrical potential expressed in volts. Thus, an electron experiencing a one-volt potential will have an interaction energy,

$$E = e\,(1.0\text{ V}) = 1.6021892 \times 10^{-19}\text{ C-V} = 1.6021892 \times 10^{-19}\text{ J}$$

This quantity serves as a measure for a special unit of energy called the electron volt (eV). Simply, 1 eV is defined to be $1.6021892 \times 10^{-19}$ J.

The cgs unit of magnetic induction is the gauss (G). In the SI system, the unit is the tesla (T), and the relation is 1 G = $10^{-4}$ T.

*Atomic Units.* A special system of units called atomic units (a.u.) proves particularly convenient in the quantum mechanics of the electronic structure of atoms and molecules. This is the system of units where the measures of mass, length, and time are chosen such that three fundamental constants, $m_e$, $\hbar$, and e, take on values of exactly 1.0. This may be accomplished because we specify a system of units by three measures, and this choice presents no more than three constraints on those measures. The resulting atomic unit of length is called the bohr (after Niels Bohr) and the atomic unit of energy is called the hartree (after D. R. Hartree).

To illustrate the use of atomic units, let us calculate the ground state energy of the hydrogen atom (under the assumption of an infinitely massive nucleus), which is known to be

$$E = -m_e e^4 / (2\hbar^2) = -2.18 \times 10^{-11}\text{ erg}$$

In atomic units, however, we have that E = $-1/2$ a.u. Therefore, the conversion between atomic units of energy and ergs is

$$1.0\text{ hartree (h)} = 2\,(2.18 \times 10^{-11}\text{ erg}) = 4.36 \times 10^{-11}\text{ erg}$$

Other energy conversions are given in Table V.1.

TABLE V.1 Energy Conversion Factors[a].

| | | erg | $cm^{-1}$ | eV | kcal/mol |
|---|---|---|---|---|---|
| 1.0 erg | = | 1.0 | $5.0340 \times 10^{15}$ | $6.2415 \times 10^{11}$ | $1.4383 \times 10^{13}$ |
| 1.0 joule | = | $10^{7}$ | $5.0340 \times 10^{22}$ | $6.2415 \times 10^{18}$ | $1.4383 \times 10^{20}$ |
| 1.0 $cm^{-1}$ | = | $1.9865 \times 10^{-16}$ | 1.0 | $1.2399 \times 10^{-4}$ | $2.8672 \times 10^{-3}$ |
| 1.0 eV | = | $1.6022 \times 10^{-12}$ | 8065.2 | 1.0 | 23.045 |
| 1.0 h | = | $4.3598 \times 10^{-11}$ | $2.1947 \times 10^{5}$ | 27.212 | 627.1 |

[a] The conversion to kcal/mol means the number of kcal for one mole of substance if each atom or molecule had the given energy in the left column. The relatively large energy units of ergs and joules are often used as ergs/mol or joules/mol, and so the conversion factors would be those shown divided by Avogadro's constant. Thus, 1.0 joule/mol = $1.4383 \times 10^{20}/6.02205 \times 10^{23} = 0.239 \times 10^{-3}$ kcal/mol.

*The Electromagnetic Spectrum.* The electromagnetic spectrum has regions that are referred to by names that have arisen, in some cases, because of different instrumental techniques. Table V.2 lists the general designations and gives rough cutoff values for the ranges.

One unit of length that is still widely used for wavelengths is the ångstrom, abbreviated Å. The conversion is 1.0 Å = $10^{-8}$ cm.

TABLE V.2 The Electromagnetic Spectrum.

| Radiation Region | Wavelengths $\lambda$ (cm) | Frequencies $\nu = c/\lambda$ ($sec^{-1}$) | Photon Energy $1/\lambda = \nu/c$ ($cm^{-1}$) |
|---|---|---|---|
| Radio | 10 to $10^{6}$ | $3 \times 10^{4}$ to $3 \times 10^{9}$ | $10^{-6}$ to 0.1 |
| Microwave | 1 to 10 | $3 \times 10^{9}$ to $3 \times 10^{10}$ | 0.1 to 1 |
| Infrared | $10^{-4}$ to 1 | $3 \times 10^{10}$ to $3 \times 10^{14}$ | 1 to $10^{4}$ |
| Visible | $4 \times 10^{-5}$ to $10^{-4}$ | $3 \times 10^{14}$ to $7.5 \times 10^{14}$ | $10^{4}$ to $2.5 \times 10^{4}$ |
| Ultraviolet | $10^{-7}$ to $4 \times 10^{-5}$ | $7.5 \times 10^{14}$ to $3 \times 10^{17}$ | $2.5 \times 10^{4}$ to $10^{7}$ |
| X-ray | $< 10^{-7}$ | $> 3 \times 10^{17}$ | $10^{7}$ |

TABLE V.3 Values of Constants.

| Constant | Symbol | Value |
|---|---|---|
| Speed of light | c | $2.997925 \times 10^{10}$ cm/sec |
| Electron charge | e | $1.602189 \times 10^{-20}$ emu |
| | | $1.602189 \times 10^{-20}$ $cm^{1/2}$ $g^{1/2}$ |
| | | $1.602189 \times 10^{-19}$ C |
| Planck's constant | $\hbar$ | $1.054589 \times 10^{-27}$ erg-sec |
| | h | $6.626176 \times 10^{-27}$ erg-sec |
| Boltzmann's constant | k | $1.380662 \times 10^{-16}$ erg / K |
| | | 0.695031 $cm^{-1}$ / K |
| Mass of an electron | $m_e$ | $9.109534 \times 10^{-28}$ g |
| Mass of a proton | $m_p$ | $1.672649 \times 10^{-24}$ g |
| Bohr magneton | $\mu_B$ | $9.274078 \times 10^{-21}$ erg / G |
| Nuclear magneton | $\mu_o$ | $5.050824 \times 10^{-24}$ erg / G |
| Avogadro's constant | $N$ | $6.02205 \times 10^{23}$ |

# *Solutions to Selected Chapter Exercises*

2-1. $H(x, y, z, p_x, p_y, p_z) = T + V$

$$= (p_x^2 + p_y^2 + p_z^2) / 2m + mgz$$

$$\frac{\partial H}{\partial q_i} = -\dot{p}_i \quad \text{gives} \quad \dot{p}_x = 0 \quad \dot{p}_y = 0 \quad \dot{p}_z = -mg$$

$$\frac{\partial H}{\partial p_i} = \dot{q}_i \quad \text{gives} \quad p_x / m = \dot{x} \quad p_y / m = \dot{y} \quad p_z / m = \dot{z}$$

2-3. $k = 2c$

$x_o = -b / 2c$

$V_o = a - b^2 / 4c$

2-5. From the given relation,

$$Ae^{-i\omega t} + Be^{i\omega t} = A\{\cos(-\omega t) + i\sin(-\omega t)\} + B\{\cos(\omega t) + i\sin(\omega t)\}$$

$$= [A + B]\cos(\omega t) + i[B - A]\sin(\omega t)$$

For Eqs. (2-8a) and (2-8b) to be the same function, it must be that

$A + B = b$ and $i[B - A] = a$.

If $b = 0$, it would mean $A = -B$. And then if $a = 1$, it would mean that $i[B - (-B)] = 1$, or that $B = -i/2$. For the general case with $b = 0$, then $B = -ia/2$ and $A = ia/2$.

2-9. In Cartesian coordinates, the Hamiltonian for the single particle is

$$H = \frac{1}{2m}(p_x^2 + p_y^2 + p_z^2) + \frac{1}{2}k(x^2 + y^2 + z^2)$$

and if this is transformed to spherical polar coordinates, the potential energy in H simplifies to $kr^2/2$. Using relations similar to those between x, y, z and r, θ, φ (only letting $x = x_2 - x_1$, and so on for y and z) and the relation of their time derivatives, the translational energy in sperical polar coordinates is

$$T = \frac{1}{2m}(\dot{r}^2 + r^2\dot{\theta}^2 + r^2\sin^2\theta\,\dot{\phi}^2)$$

This expression is also the translational motion due to vibrational and rotational motion for the system of two particles. So this system of one particle is equivalent to the two particle system where the center of mass translational energy is zero (X, Y, Z are constant).

3-1. Electron's momentum is $p = 0.1 \times 3.00 \times 10^{10}$ cm/sec $\times 9.11 \times 10^{-28}$ g

$= 2.73 \times 10^{-18}$ g-cm/sec

De Broglie wavelength is $\lambda = h/p = 6.63 \times 10^{-27} / 2.73 \times 10^{-18}$ cm

$= 2.43 \times 10^{-9}$ cm

For a marble to have the same wavelength, it must have the same momentum. Thus,

$$(10.0\text{ g})(v) = 2.73 \times 10^{-18}\text{ g-cm/sec}$$

$$v = 2.73 \times 10^{-19}\text{ cm/sec}$$

3-2. *b*. The commutator is evaluated by applying it to an arbitrary function.

$$[\hat{B}^2, \hat{A}]f(x) = [\frac{d^2}{dx^2}, x]f(x) = \frac{d^2}{dx^2}(x f(x)) - x\frac{d^2}{dx^2}f(x)$$

$$= \left(2\frac{d}{dx}f(x) + x\frac{d^2}{dx^2}f(x)\right) - x\frac{d^2}{dx^2}f(x) = 2\frac{d}{dx}f(x)$$

Therefore, we have that $[\hat{B}^2, \hat{A}] = 2\,d/dx$ since the net effect of operating with this commutator on any function will yield the same result as operating on the function with d/dx and then multiplying by the number 2.

3-7. We can verify that $\hat{A} = \hat{H}\hat{H}$, and from that we can show that $\psi_0$ is an eigenfunction: $\hat{A}\psi_0 = \hat{H}\,(\hat{H}\psi_0) = \hat{H}(E_0\,\psi_0) = E_0\,\hat{H}\psi_0 = E_0^2\,\psi_0$.

3-8. $<p^2>_{n=0} \;=\; \int_{-\infty}^{\infty} \psi_0{}^*(x)\,\hat{p}^2\,\psi_0(x)\,dx$, and likewise for the n = 1 state.

To carry out the evaluation of this integral the momentum-squared operator must be applied to the wavefunction. We can make use of what was already done in the course of applying the Hamiltonian to $\psi_0$.

$$-\frac{\hbar^2}{2m}\frac{d^2}{dx^2}\psi_0(x) \;=\; \left[-\frac{\hbar^2}{2m}\beta^4 x^2 + \frac{\hbar^2\beta^2}{2m}\right]\psi_0(x)$$

$$<p^2>_{n=0} \;=\; -\hbar^2\beta^4\int_{-\infty}^{\infty}\psi_0{}^*(x)\,x^2\,\psi_0(x)\,dx \;+\; \hbar^2\beta^2\int_{-\infty}^{\infty}\psi_0{}^*(x)\,\psi_0(x)\,dx$$

The second of the two integrals is equal to one because of the normalization condition on the wavefunction. The first integral requires consulting an integrals table or other means. But first, it is helpful to substitute with the variable z.

$$-\hbar^2\beta^4\int_{-\infty}^{\infty}\left(\frac{\beta^2}{\pi}\right)^{1/4} e^{-\beta^2x^2/2}\,x^2\left(\frac{\beta^2}{\pi}\right)^{1/4} e^{-\beta^2x^2/2}\,dx$$

$$=\; -\frac{\hbar^2\beta^2}{\sqrt{\pi}}\int_{-\infty}^{\infty} e^{-z^2/2}\,z^2\,e^{-z^2/2}\,dz \;=\; -\frac{\hbar^2\beta^2}{\sqrt{\pi}}\int_{-\infty}^{\infty} z^2\,e^{-z^2}\,dz$$

And from the integral tables in Appendix III and the realization that the integral will be the same value for the integration range 0 to $\infty$ as for $-\infty$ to 0,

$$=\; -\frac{\hbar^2\beta^2}{\sqrt{\pi}}\,\frac{\sqrt{\pi}}{2}$$

So, combining the intermediate values,

$$<p^2>_{n=0} = -\hbar^2\beta^2/2 + \hbar^2\beta^2 = \hbar^2\beta^2/2$$

3-10. The uncertainty can be evaluated from the expectation value of the momentum squared and the expectation value of the momentum for a particular state. For the n = 0 state, the expectation value $<p>$ must be identically zero because the integral is over a function that is overall odd with respect to x = 0. From Prob. 3-8, we have a value for $<p^2>$, and so

$$\Delta p_{n=0} = \sqrt{<p^2>_{n=0} - <p>^2_{n=0}} = \sqrt{\hbar^2\beta^2/2} = \hbar\beta/\sqrt{2}$$

3-16. $<x>_n = l/2$ and $\Delta x_n = l\sqrt{\left(\frac{1}{3} - \frac{1}{2(n\pi)^2}\right) - \frac{1}{4}}$ . Thus, for example, if $l = \pi$, then $\Delta x_1 = 0.56786$, $\Delta x_2 = 0.83514$, and $\Delta x_3 = 0.87573$.

3-20. $l$ = 9.54Å; n = 4.

4-2. The entire set of wavefunctions would be eigenfunctions of the x operator or of the $p_x$ operator only if they were to commute with the Hamiltonian.

$$[x, H]\,f(x) = [x, \frac{1}{2}kx^2]\,f(x) + [x, -\frac{\hbar^2}{2m}\frac{d^2}{dx^2}]\,f(x)$$

$$= \left(\frac{1}{2}kx^3 - \frac{1}{2}kx^3\right)f(x) - x\frac{\hbar^2}{2m}\frac{d^2}{dx^2}f(x) + \frac{\hbar^2}{2m}\frac{d^2}{dx^2}(x\,f(x))$$

$$= \frac{\hbar^2}{m}\frac{d}{dx}f(x) \neq 0$$

Thus, x does not commute with the Hamiltonian. It can be shown that $p_x$ does not commute with H either. So, the wavefunctions are not eigenfunctions of either of these operators.

4-5. The energy level expression for an isotropic three-dimensional oscillator is

$$E_{n_x n_y n_z} = (n_x + \frac{1}{2})\hbar\omega + (n_y + \frac{1}{2})\hbar\omega + (n_z + \frac{1}{2})\hbar\omega$$

This results because of separability in the three directions and because the isotropic character of the problem means $k_x = k_y = k_z$, which in turn means the frequencies for vibration in the x-, y-, and z-directions are the same. So, the energy expression may be written simply as

$$E_{n_x n_y n_z} = \left( n_x + n_y + n_z + \frac{3}{2} \right) \hbar \omega$$

To determine the degeneracy of an energy level we must find all combinations of the three quantum numbers that give the same energy, remembering that for a harmonic oscillator the quantum numbers may be zero or any positive integer. From the energy expression, it should be clear that it is just the sum of the three quantum numbers that is important, and we will name this sum N; i.e., $N = n_x + n_y + n_z$. As N increases, the energy increases. The smallest allowed value for N is zero, and then all three of the quantum numbers are zero. This level is not degenerate because there is no other choice of the quantum numbers which gives N = 0. For the first excited state energy level, which means N = 1, there are three different choices of the quantum numbers: $n_x$ could be 1, or $n_y$ could be 1, or $n_z$ could be 1, while the other two are zero. Thus, there are three different states with this energy; the degeneracy of the second level (first excited level) is three. The following list continues this process.

| | $(n_x, n_y, n_z)$ of a state with the given N: |
|---|---|
| N = 0 | (0,0,0) degeneracy = 1 (nondegenerate) |
| N = 1 | (1,0,0), (0,1,0), (0,0,1) degeneracy = 3 |
| N = 2 | (2,0,0), (0,2,0), (0,0,2), (1,1,0), (1,0,1), (0,1,1) degeneracy = 6 |
| N = 3 | (3,0,0), (2,1,0), (2,0,1), (1,2,0), (1,1,1), (1,0,2), (0,3,0), (0,2,1), (0,1,2), (0,0,3) degeneracy = 10 |

4-6. It is necessary to examine the energies of the states systematically. With n = 1, the first term in this energy expression is at its lowest value. So, we may start there and find all the energies for the possible values of m. Then, we may repeat this for n = 2, and so on.

n = 1 m = 0 $E/\hbar a = -1$

| | | |
|---|---|---|
| $n = 1$ | $m = 1$ or $-1$ | $E/\hbar a = -3/4$ |
| $n = 1$ | $m = 2$ or $-2$ | $E/\hbar a = 0$ |
| $n = 2$ | $m = 0$ | $E/\hbar a = -1/4$ |
| $n = 2$ | $m = 1$ or $-1$ | $E/\hbar a = 0$ |
| $n = 2$ | $m = 2$ or $-2$ | $E/\hbar a = 3/4$ |
| $n = 2$ | $m = 3$ or $-3$ | $E/\hbar a = 2$ |
| $n = 3$ | $m = 0$ | $E/\hbar a = -1/9$ |
| $n = 3$ | $m = 1$ or $-1$ | $E/\hbar a = 5/36$ |
| $n = 3$ | $m = 2$ or $-2$ | $E/\hbar a = 8/9$ |
| $n = 3$ | $m = 3$ or $-3$ | $E/\hbar a = 77/36$ |
| $n = 3$ | $m = 4$ or $-4$ | $E/\hbar a = 35/9$ |

There are 21 different states because there are 21 different allowed values for the two quantum numbers. From the energy level expression we obtain 11 different energy values, and so there are 11 energy levels. The degeneracies are obtained from counting up the possibilities that yielded each particular energy value. The lowest level or the ground state is not degenerate with any other. The other two m = 0 states are nondegenerate, too (degeneracy = 1). There are 4 states with energy of zero, and so this level has a fourfold degeneracy. The remaining seven levels are doubly degnerate (degeneracy = 2). If we add up the degeneracies, $3 \times 1 + 4 + 7 \times 2$, we get 21, the number of states.

*4-15.* $\hat{L}^2 (\sin^m\theta \cos\theta\, e^{im\phi})$

$$= -\hbar^2 \left( \frac{1}{\sin\theta}\frac{\partial}{\partial\theta}\sin\theta\frac{\partial}{\partial\theta} + \frac{1}{\sin^2\theta}\frac{\partial^2}{\partial\phi^2} \right)(\sin^m\theta\ \cos\theta\, e^{im\phi})$$

$$= -\hbar^2\left[\, m^2 \sin^{m-2}\theta \cos^3\theta \;-\; (3m+2)\sin^m\theta \cos\theta \,\right] e^{im\phi}$$

$$-\hbar^2 (-m^2) \sin^{m-2}\theta \cos\theta\, e^{im\phi}$$

with the term in [ ] from the differentiation with respect to $\theta$ and the last term from the differentiation with respect to $\phi$. If the last term is multiplied by $1 = \sin^2\theta + \cos^2\theta$, then it may be combined with the term in [ ], yielding

$$= \hbar^2 (m^2 + 3m + 2) \sin^m\theta \cos\theta\, e^{i m\phi}$$

This shows that the function is indeed an eigenfunction of $\hat{L}^2$. The eigenvalue is $\hbar^2 (m^2 + 3m + 2)$, or $\hbar^2 (m + 1)(m + 2)$.

*4-18.* $Y_{20}(\theta,\phi) = \Theta_{20}(\theta)\,\Phi_0(\phi) = \frac{1}{4}\sqrt{\frac{5}{\pi}}\left(3\cos^2\theta - 1\right)$

$$\hat{L}_x Y_{20} = -i\hbar \frac{3}{2}\sqrt{\frac{5}{\pi}} \cos\theta \sin\theta \sin\phi$$

$$\left\langle Y_{20} \mid \hat{L}_x Y_{20}\right\rangle = \left\langle \text{integral over } \theta\right\rangle \int_0^{2\pi} \sin\phi \, d\phi = 0$$

$$\hat{L}_x (\hat{L}_x Y_{20}) = -\hbar^2 \frac{3}{2}\sqrt{\frac{5}{\pi}} \left(\sin^2\phi \sin^2\theta - \cos^2\theta\right)$$

$$\left\langle Y_{20} \mid \hat{L}_x^2 Y_{20}\right\rangle = 0$$

*4-20.* Couple the momenta two at a time. Coupling $j_1$ and $j_2$ results in values of $j_a = 3, 2, 1$. Then, coupling each possible value of $j_a$ with $j_3$ leads to possible values of J = 4, 3, 2, 3, 2, 1, 2, 1, 0.

*4-21.* Two sources may combine such that the total angular momentum quantum number, J, ranges (in steps of one) from the sum to the absolute value of the difference, according to the rule of angular momentum addition. With two hypothetical soucres with quantum numbers of 1/2, the sum is one and the difference is zero. Therefore, 0 and 1 are the possible values for the quantum number of the total angular momentum.

If there is a third source, it adds to the resultant of the first two. Thus, $J^{\text{3-sources}} = \frac{1}{2} + 0 = \frac{1}{2}$ because of the zero sum of the first two sources, and $J^{\text{3-sources}} = \frac{1}{2} + 1, \ldots \left|\frac{1}{2} - 1\right| = \frac{3}{2}$ and $\frac{1}{2}$ because of the case where the sum from the first two sources is one. We interpret this to mean that the three sources may combine to give J = 3/2 or they may combine in two *different* ways to give J = 1/2.

The fourth source may combine with the 3/2 resultant from three sources to give $J^{\text{4-sources}} = 2$ or 1. It may combine with either of the 1/2 resultants to give $J^{\text{4-sources}} = 1$ or 0. This may be expressed concisely as

$$J^{\text{4-sources}} = 2, 1, 1, 1, 0, 0$$

The allowed possibilities for the resultant J quantum number with the four sources is this list of integers. Integers are repeated if there are different ways that they may arise.

4-23. $E_n^{(1)} = \langle H^{(1)} \rangle = \langle \psi_n^{(0)} \mid V(x)\, \psi_n^{(0)} \rangle = a \langle \psi_n^{(0)} \mid x\, \psi_n^{(0)} \rangle = \frac{a\,l}{2}$

$$\Psi_n^{(1)} = \sum_j \frac{-16ml^3\, a\, nj}{\pi^4 \hbar^2 (n-j)^3 (n+j)^3}\, \Psi_j^{(0)}, \; j + n = \text{odd}$$

$$E_n^{(2)} = \sum_j \frac{128\, m\, l^4 a^2 n^2 j^2}{\pi^6 \hbar^2 (n-j)^5 (n+j)^5}, \; j + n = \text{odd}$$

5-3. Using the definitions of $Y_{JM}$ and the recursion relationship for the associated Legendre polynomials,

$$\cos\theta\, Y_{J'M'} = \sqrt{\frac{(2J'+1)(J'-|M'|)!}{2\,(J'+|M'|)!}}\; \Phi_M(\phi) \cos\theta\, P_{J'}^{|M'|}(\cos\theta)$$

$$= \sqrt{\frac{(2J'+1)(J'-|M'|)!}{2\,(J'+|M'|)!}}\; \Phi_M(\phi) \frac{1}{2J'+1} \left( (J'+1-|M'|)\, P_{J'+1}^{|M'|} + (J'+|M'|)\, P_{J'-1}^{|M'|} \right)$$

$$= \frac{1}{2J'+1} \left( \sqrt{\frac{(2J'+1)\left[(J'+1)^2 - |M'|^2\right]}{2J'+3}}\; Y_{J'+1\,M'} + \sqrt{\frac{(2J'+1)\left(J'^2 - |M'|^2\right)}{2J'-1}}\; Y_{J'-1\,M'} \right)$$

With this result, $\langle Y_{JM} \mid \cos\theta\, Y_{J'M'} \rangle$ reduces to overlap integrals of the spherical harmonics, and from their orthogonality relationships Eq. (5-27) is obtained.

5-5. $E_{n+1,n} = E_{n+1} - E_n = \omega_e - 2(n+1)\,\omega_e\chi_e + \{3\,[(n+\frac{1}{2})^2 + (n+\frac{1}{2})\,] + 1\}\omega_e y_e$. Using the given transition frequencies to set up three equations, the three unknowns can be solved for and then $\omega_e =$

1749.92 $cm^{-1}$, $\omega_e\chi_e = 87.50$ $cm^{-1}$, and $\omega_e y_e = 8.33$ $cm^{-1}$. On the other hand, letting $\omega_e y_e = 0$ and setting up two equations in two unknowns leads to $\omega_e = 1700$ $cm^{-1}$, $\omega_e\chi_e = 50$ $cm^{-1}$. Alternatively, including all three transitions and fitting them by a least squares analysis with the two term formula leads to $\omega_e = 1666.7$ $cm^{-1}$ and $\omega_e\chi_e = 37.5$ $cm^{-1}$.

5-8. $r_e = 1.596$ Å.

| J | ΔE |
|---|---|
| 0-1 | 1036.77 ($cm^{-1}$) |
| 1-2 | 1044.81 |
| 2-3 | 1052.72 |
| 3-4 | 1060.55 |
| 4-5 | 1068.36 |
| 5-6 | 1076.22 |
| 6-7 | 1084.21 |

5-13. In the course of the first two stretching modes, the left-right symmetry of the molecule is preserved and so the dipole always remains zero. This means the dipole is unchanging, and so these two modes are infrared-inactive. In the third stretching mode, the symmetry is not preserved. A dipole along the molecular axis will grow in as the molecule vibrates away from its equilibrium structure. The dipole changes in the course of this vibration, and so this mode is infrared-active. The first bending mode is infrared-active because a dipole will develop that is perpendicular to the molecular axis. The second bending mode preserves the symmetry in such a way that the dipole will remain at zero. It is not active.

6-1. $L_3^0 = e^z \left(\frac{d}{dz}\right)^3 z^3 e^{-z} = 6 - 18z + 9z^2 - z^3 \quad L_3^2 = \left(\frac{d}{dz}\right)^2 L_3^0 = 18 - 6z$

6-6. $E_{ionization} = (Z_{ion} / Z_H)^2 E_H = Z_{ion}^2 E_H$.

| Ion | $E_{ionization}$ ($cm^{-1}$) |
|---|---|
| $He^+$ | 438,960 |
| $Li^{2+}$ | 987,660 |
| $C^{5+}$ | 3,950,640 |
| $Ne^{9+}$ | 10,974,000 |

6-8. Since each shell is filled in a $s^2$ $p^6$ $d^{10}$ occupancy, there is only one entry in the table, with $M_L = 0$ and $M_S = 0$; therefore L = 0 and S = 0. For an occupancy of $1s^2$ $2p^1$, the table is

| $1s_0$ | $2p_1$ | $2p_0$ | $2p_{-1}$ | $M_L$ | $M_S$ |
|---|---|---|---|---|---|
| ↑↓ | ↑ | | | 1 | 1/2 |
| ↑↓ | ↓ | | | 1 | −1/2 |
| ↑↓ | | ↑ | | 0 | 1/2 |
| ↑↓ | | ↓ | | 0 | −1/2 |
| ↑↓ | | | ↑ | −1 | 1/2 |
| ↑↓ | | | ↓ | −1 | −1/2 |

which leads to the term symbol $^2P$. Since the s shell is full, it has only one possible occupancy, for which $M_L = 0$ and $M_S = 0$. So this table (without the electrons of the s orbital) is correct for a $2p^1$ occupancy, and the term symbol $^2P$ must be correct for a $2p^1$ occupancy as well.

6-10. $^1D_2$, $^3P_{2,1,0}$, and $^1S_0$.

6-12. $^4S_{3/2}$.

6-18. The ground state of carbon is $^3P$. The excited states are $^3D$, $^1D$, $^3P$, $^1P$, $^5S$, and $^3S$. Following the selection rules $\Delta S = 0$ and $\Delta L = 1$, the allowed transitions are to $^3D$ and $^3S$.

6-22. The number of particles allowed per orbital equals the number of different spin states, so for s = 3/2, the exclusion principle would allow four particles [2 (3/2) + 1].

6-26. LiH, 1; $Be_2$, 0; $N_2$, 3; $F_2$, 1; $NeF^+$, 1.

7-5.

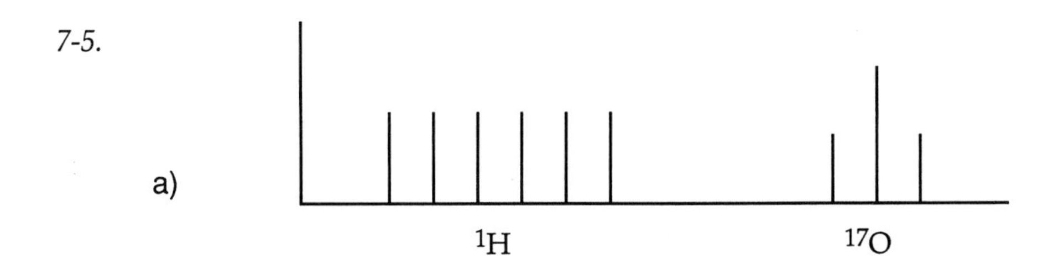

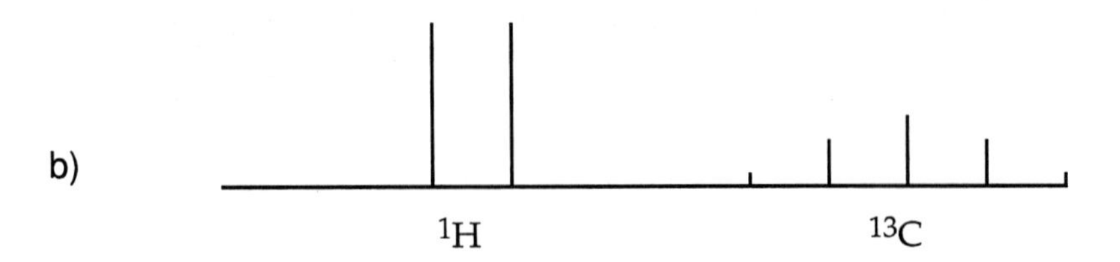


*7-7.* One nucleus, 1 : 1 : 1
Two nuclei, 1 : 2 : 3 : 2 : 1
Three nuclei, 1 : 3 : 6 : 7 : 6 : 3 : 1

*7-11.*

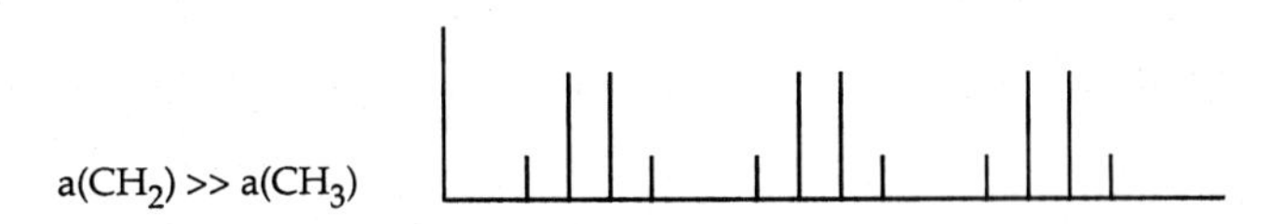


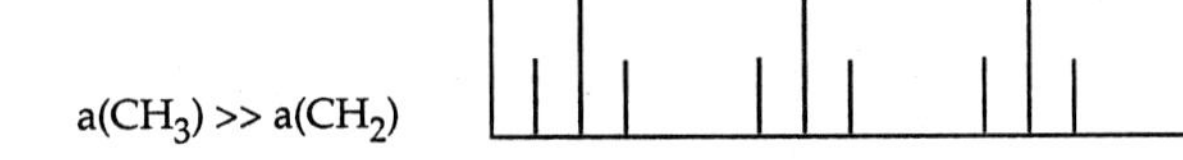


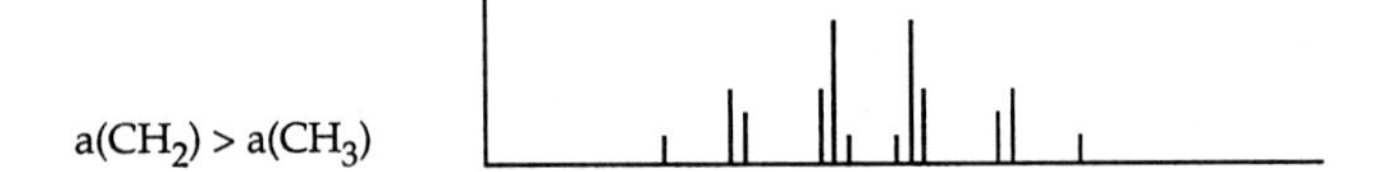

# *Subject Index*